Texts in Computer Science

Series editors

David Gries, Department of Computer Science, Cornell University, Ithaca, NY, USA

Orit Hazzan, Faculty of Education in Technology and Science, Technion—Israel Institute of Technology, Haifa, Israel

Fred B. Schneider, Department of Computer Science, Cornell University, Ithaca, NY, USA

More information about this series at http://www.springer.com/series/3191

Peter R. Turner · Thomas Arildsen
Kathleen Kavanagh

Applied Scientific Computing

With Python

 Springer

Peter R. Turner
Clarkson University
Potsdam, NY
USA

Kathleen Kavanagh
Clarkson University
Potsdam, NY
USA

Thomas Arildsen ⓘ
Aalborg University
Aalborg
Denmark

ISSN 1868-0941   ISSN 1868-095X   (electronic)
Texts in Computer Science
ISBN 978-3-319-89574-1   ISBN 978-3-319-89575-8   (eBook)
https://doi.org/10.1007/978-3-319-89575-8

Library of Congress Control Number: 2018937706

Printed on acid-free paper

This Springer imprint is published by the registered company Springer International Publishing AG part of Springer Nature
The registered company address is: Gewerbestrasse 11, 6330 Cham, Switzerland

Preface

This book represents a modern approach to teaching numerical methods—or scientific computing—to students with a broad range of primary interests. The underlying mathematical content is not new, of course, but the focus on applications and models (or modeling) is. Today's mathematics or computer science students have a strong desire to see the relevance of what they are studying in a practical way. This is even more true for students of other STEM disciplines in their mathematics and computing classes.

In an introductory text such as this, it is difficult to give complete detail on the application of the methods to "real-world" engineering, or economics, or physical science, or social science applications but it is important to connect those dots. Throughout the book, we emphasize applications and include opportunities for both problem- and project-based learning through examples, exercises, and projects drawn from practical applications. Thus, the book provides a self-contained answer to the common question "why do I need to learn this?"

That is a question which we believe any student has a right both to ask and to expect a reasonable and credible answer, hence our focus on Applied Scientific Computing. The intention is that this book is suitable for an introductory first course in scientific computing for students across a range of major fields of study. Therefore, we make no pretense at a fully rigorous development of all the state-of-the-art methods and their analyses. The intention is to give a basic understanding of the need for, and methods of, scientific computing for different types of problems. In all cases, there is sufficient mathematical justification and practical evidence to provide motivation for the reader.

Any text such as this needs to provide practical guidance on coding the methods, too. Coding a linear system solver, for example, certainly helps the student's understanding of the practical issues, and limitations, of its use. Typically students find introductory numerical methods more difficult than their professors expect. Part of the reason is that students are expected to combine skills from different parts of their experience—different branches of mathematics, programming and other computer science skills, and some insight in applications fields. Arguably, these are independent skill sets and so the probabilities of success multiply.

Our choice of Python for the computer language is based on the desire to minimize the overhead of learning a high-level language or of the intricacies (and cost) of a specific application package. The coding examples here are intended to be relatively easily readable; they are not intended to be professional-level software. The Python code is there to facilitate the learning and understanding of the basic material, rather than being an objective in itself.

Turning briefly to the content of the book, we gave considerable thought to the ordering of material and believe that the order we have is one good way to organize a course, though of course every instructor will have his/her own ideas on that. The chapters are largely independent and so using them in a different order will not be problematic.

We begin with a brief background chapter that simply introduces the main topics application and modeling, Python programming, sources of error. The latter is exemplified at this point with simple series expansions which should be familiar from calculus but need nothing more. These series expansions also demonstrate that the "most obvious" approach that a beginning student might adopt will not always be practical. As such it serves to motivate the need for more careful analysis of problems and their solutions. Chapter 2 is still somewhat foundational with its focus on number representation and errors. The impact of how numbers are represented in a computer, and the effects of rounding and truncation errors recur in discussing almost any computational solution to any real-life problem.

From Chap. 3 onwards, we are more focused on modeling, applications, and the numerical methods needed to solve them. In Chap. 3 itself the focus is on numerical calculus. We put this before interpolation to reflect the students' familiarity with the basic concepts and likely exposure to at least some simple approaches. These are treated without any explicit reference to finite difference schemes.

Chapters 4 and 5 are devoted to linear and then nonlinear equations. Systems of linear equations are another topic with which students are somewhat familiar, at least in low dimension, and we build on that knowledge to develop more robust methods. The nonlinear equation chapter also builds on prior knowledge and is motivated by practical applications. It concludes with a brief treatment of the multivariable Newton's method.

The final two chapters are on interpolation and the numerical solution of differential equations. Polynomial interpolation is based mostly on a divided difference approach which then leads naturally to splines. Differential equations start from Euler's method, and then Runge Kutta, for initial value problems. Shooting methods for two-point boundary value problems are also covered, tying in many ideas from the previous chapters.

All of this material is presented in a gentle way that relies very heavily on applications and includes working Python code for many of the methods. The intention is to enable students to overcome the combined demands of the mathematics, the computing, the applications, and motivation to gain a good working insight into the fundamental issues of scientific computing. The impact of the inevitable reliance on algebraic manipulation is largely minimized through careful explanation of those details to help the student appreciate the essential material.

All of us have benefited from many helpful discussions both on the philosophy and the details of teaching this content over many years–more for some than others! Those influencers are too many to list, and so we simply thank them en masse for all the helpful conversations we have had over the years.

Potsdam, USA   Peter R. Turner
Aalborg, Denmark   Thomas Arildsen
Potsdam, USA   Kathleen Kavanagh

Contents

Motivation and Background

The need for a workforce with interdisciplinary problem solving skills is critical and at the heart of that lies applied scientific computing. The integration of computer programming, mathematics, numerical methods, and modeling can be combined to address global issues in health, finances, natural resources, and a wide range of complex systems across scientific disciplines. Those types of issues all share a unique property–they are comprised of open-ended questions. There is not one necessarily right answer to the question, "How should we manage natural resources?" As a matter of fact, that question in and of itself needs to be better defined, but we know it is an issue. Certainly "a solution" requires the use of mathematics, most likely through the creation, application and refinement of innovative mathematical models that represent the physical situation. To solve those models, requires computer programming, efficient and accurate simulation tools, and likely the analysis and incorporation of large data sets.

In this chapter, we motivate the need for applied scientific computing and provide some of the foundational ideas used throughout the rest of the text. We proceed by describing mathematical modeling which is our primary tool for developing the mathematical problems that we need scientific computing to address. Then, we point to how computational science is at the heart of solving real-world problems, and finally we review some important mathematical ideas needed throughout.

1.1 Mathematical Modeling and Applications

To get an idea of what math modeling is all about, consider the following questions; (1) *A new strain of the flu has surfaced. How significant is the outbreak?* and (2) *A sick person infects two people/day. If your dorm consists 100 people, and two people are initially sick, how long before everyone is infected?* When you read the first question, you might not even think of it as a math question at all. Arguably, it is not–but insight could be gained by using mathematics (and ultimately scientific

© Springer International Publishing AG, part of Springer Nature 2018
P. R. Turner et al., *Applied Scientific Computing*, Texts in Computer Science,
https://doi.org/10.1007/978-3-319-89575-8_1

computing). When you read the second question, you probably feel more confident and could get started immediately doing some computations. You might begin by making a table of values and calculating how many people are sick each day until you reach 100. The answer would then be the day that happened. You could conceivably, in a few lines of computer code, write a program to solve this problem even for a more general case where the population size and the rate of spreading are inputs to your code.

Both questions have the same context, understanding how a disease spreads through a population. What is the main difference between these questions? The second one provided you with everything you need to get to the answer. If your entire class approached this problem, you would all get the same answer.

With the first question though, you initially need to decide what is even meant by "significant" and then think about how to quantify that. Modeling allows us to use mathematics to analyze an outbreak and decide if there is cause for alarm or to propose an intervention strategy. Math modeling allows for interpretation and innovation and there is often not an obvious single solution. If your entire class split up and worked on the first question, you would likely see a wide range of approaches. You might make an assumption along the lines of; *The rate at which a disease spreads is proportional to product of the people who have it and the people who do not*. In this case, you may ultimately wind up with a system of differential equations that model how the infected and uninfected populations change over time. Depending on the level of complexity you chose to include, numerical techniques would be needed to approximate a solution to the differential equation model.

Specifically, **a mathematical model** is often defined as a representation of a system or scenario that is used to gain qualitative and/or quantitative understanding of some real world problem, and to predict future behavior. Models are used across all disciplines and help people make informed decisions. Throughout this text, we will use real-world applications like the second question to motivate and demonstrate the usefulness of numerical methods. These will often be posed as mathematical models that have already been developed for you, and it will be your role to apply scientific computing to get an answer. In addition to that, we provide math modeling in which numerical methods have powerful potential to give insight into a real-world problem but you first need to develop the model. We provide a brief overview of the math modeling process below for inspiration.

Overview of the Math Modeling Process

Math modeling is an iterative process and the key components may be revisited multiple times and not necessarily even done in order, as we show in Fig. 1.1. We review the components here and point the reader towards the free resource *Math Modeling: Getting Started Getting Solutions, Karen M. Bliss et al., SIAM, Philadelphia, 2013* for more information, examples, and practice.

- **Defining the Problem Statement** If the initial idea behind the problem is broad or vague, you may need to form a concise problem statement which specifies what the output of your model will be.

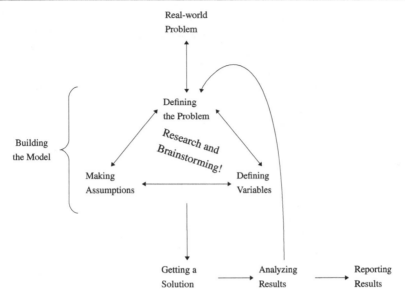

Fig. 1.1 The math modeling process

- **Making Assumptions** It is necessary to make assumptions to help simplify the problem and decide which factors are important to include in your model.
- **Defining Variables** List the important factors as quantifiable variables with specified units. You may need to distinguish between independent variables, dependent variables, and model parameters.
- **Getting a Solution** Determining a solution may involve evaluating a function, running simulations or approaching the problem numerically, depending on the type of model you developed.
- **Analysis and Model Assessment** Analyze the results to assess the quality of the model. Does the output make sense? Is it possible to make (or at least point out) possible improvements?
- **Reporting the Results** Your model is only useful if you can communicate how to use it to others.

1.2 Applied Scientific Computing

In this text, you will learn the tools to tackle these sorts of messy real-world problems and gain experience by using them. Within that modeling process in Fig. 1.1, we will be primarily concerned with solving the resulting model and analyzing the results (although there will be opportunities to build models as well). In the process of solving real-world problems, once a mathematical representation is obtained, then one must choose the right tool to find a solution.

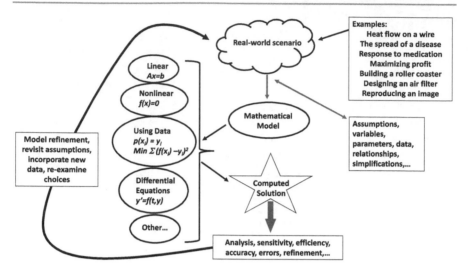

Fig. 1.2 Examples of types of problems resulting from mathematical modeling of real world problems

The mathematical model itself can take on many forms (we suggest taking an entire course in just that topic!). Figure 1.2 shows where the topics covered in this text emerge as tools to find a solution within the modeling process. Here is a brief look at the types of problems you will be able to solve after you are done with this text. First note that you learned all kinds of pencil and paper techniques for many of these "types" of problems presented throughout this book, but the purpose here is to move beyond that to solve more complex problems.

In order to be able to use scientific computing tools, it is necessary to become familiar with how numbers are stored and used in a computer. The chapter on **Number Representations and Errors** provides some background on this as well as examples of different types of errors that can impact the accuracy of your solution and tools to measure the accuracy.

In the **Numerical Calculus** chapter, you will learn tools to approximate derivatives and integrals using only function values or data. You'll also learn how to analyze the accuracy in those approximations. These can be powerful tools in practice. For example, the function you are working with may not even have an analytic integral. Sometimes you don't have a function at all. That instance may occur if a function evaluation requires output from simulation software that you used to compute some other quantity of interest. It may be the case that you only have data, for example collected from a sensor or even given to you as a graph from a collaborator, but you need to use those values to approximate an integral or a rate of change.

The chapter on **Linear Equations** is about how to solve problems of the form $Ax = b$ where A is square matrix and x and b are vectors. Linear models that lead to problems of this form arise across disciplines and are prone to errors if care is not taken in choosing the right algorithm. We discuss direct methods and iterative methods as well as ways to make the algorithmic approaches more efficient in some cases. The topic of eigenvalues is also introduced.

In Chap. 5, we address **nonlinear equations** of one variable, that is problems of the form $f(x) = 0$ and systems of nonlinear equations, where $f : R^n \rightarrow R^n$. Problems of this type almost always require iterative methods and several are presented, compared, and analyzed.

Chapter 6 on **Interpolation** covers the idea of finding a function that agrees with a data set or another function at certain points. This is different than try to fit a model as closely as possible to data (which would result in a linear or nonlinear regression or *least squares* problem and ideas from the previous two chapters). The focus in this text is on finding polynomials or piecewise polynomials that equal the function (or data) exactly.

Finally in Chap. 7, we present numerical approaches to solving **Differential Equations**, which arise routinely in math modeling, for example in the context of infectious diseases or population dynamics. Problems in that chapter have the form $y' = f(x, y)$ where here the solution is a function $y(x)$ that satisfies that differential equation. The continuous problem is discretized so that approximations at specific x values are computed. We also consider higher order differential equations and boundary value problems with an emphasis on accuracy and computational efficiency. Solution techniques here rely on ideas from the previous chapters and are a culminating way to gain appreciation for and practice with the tools presented throughout the text.

1.3 Python Programming

Throughout this text, examples are used to provide insight to the numerical methods and how they work. In some cases, a few computations, or steps through an algorithm, are shown and other times Python code is provided (along with the output). Python is an open source, interpreted, cross-platform language that was developed in the late 1980s (no compiling and it's free). The motivation was to be flexible with straightforward syntax. Python has grown to be an industry standard now and a powerful computing tool, with a wide range of libraries for mathematics, data science, modeling, visualization, and even game design. It is considered "easy" to get started with and there are a plethora of tutorials, resources, and support forums available to the scientific community. At the end of each chapter in this book, we point to Python tools and resources for further exploration. Head to
https://www.python.org
to get started. In particular, you should be sure to install the NumPy and SciPy libraries which are the standard scientific computation packages, along with Matplotlib for plotting. For an easy way to install Python and all essential packages for scientific computing, consider looking up the Anaconda Python distribution.

This book is a not a book about Python programming. For a thorough introduction to Python for scientific computing, we recommend Hans Petter Langtangen: *A Primer on Scientific Programming with Python*, Springer, 2016.

Throughout this book, we make use of the three essential scientific computing packages mentioned above. NumPy provides basic vector- and matrix-based numerical computing based on the NumPy array data structure and its accompanying functionality such as basic linear algebra, random numbers etc. SciPy provides a larger selection of higher-level functionality such as interpolation, integration, optimization etc. that works on top of NumPy. Matplotlib provides advanced plotting functionality for visualising numerical data and is used in some of the examples to illustrate the results.

In order to use additional packages such as NumPy, SciPy, and Matplotlib, they must first be imported in the Python script. This is conventionally done as:

```python
import numpy as np
import scipy as sp
import matplotlib.pyplot as plt
```

Some of the code examples in the coming chapters implicitly assume that these packages have been imported although it is not always shown in the particular example. When you come across statements in the code examples beginning with the above np., sp., or plt., assume that it refers to the above packages.

1.4 Background

In this section we provide some of the mathematical background material that is fundamental to understanding the later chapters. Many of these ideas you likely saw in Calculus I or II but now they are presented it in the context of scientific computing.

1.4.1 Series Expansions

Recall a *power series* centered about a point a has the form

$$\sum_{n=0}^{\infty} C_n(x - a)^n = C_0 + C_1(x - a) + C_2(x - a)^2 + \ldots,$$

where here C_n are coefficients and x is a variable. In its simplest definition, a power series is an infinite degree polynomial. Significant time is spent in Calculus answering the question "For what values of x will the power series converge?" and multiple approaches could be used to determine a radius and interval of convergence. Usually this is followed by an even more important question, "Which functions have a power series representation?" which lead into Taylor and MacLaurin series. Recall that the Taylor series representation of a function $f(x)$ about a point a and some radius of convergence is given by

$$f(x) = \sum_{n=0}^{\infty} \frac{f^{(n)}(a)}{n!} (x-a)^n = f(a) + f'(a)(x-a) + \frac{f''(a)}{2!}(x-a)^2 + \cdots$$

$$(1.1)$$

and a MacLaurin series was the special case with $a = 0$. This was a powerful new tool in that complicated functions could be represented as polynomials, which are easy to differentiate, integrate, and evaluate.

Two series which we will make use of are the *geometric series*

$$\frac{1}{1-x} = \sum_{k=0}^{\infty} x^k = 1 + x + x^2 + \cdots \qquad (|x| < 1) \qquad (1.2)$$

and the *exponential series*

$$\exp(x) = e^x = \sum_{k=0}^{\infty} \frac{x^k}{k!} = 1 + x + \frac{x^2}{2!} + \frac{x^3}{3!} + \cdots \qquad (\text{all } x) \qquad (1.3)$$

Other important series expansions can be derived from these, or by Taylor or MacLaurin expansions. Using the identity

$$e^{ix} = \cos x + i \sin x$$

where $i = \sqrt{-1}$, we get the following series:

$$\cos x = \sum_{k=0}^{\infty} \frac{(-1)^k x^{2k}}{(2k)!} = 1 - \frac{x^2}{2!} + \frac{x^4}{4!} \cdots \qquad (\text{all } x) \qquad (1.4)$$

$$\sin x = \sum_{k=0}^{\infty} \frac{(-1)^k x^{2k+1}}{(2k+1)!} = x - \frac{x^3}{3!} + \frac{x^5}{5!} \cdots \qquad (\text{all } x) \qquad (1.5)$$

(If you are unfamiliar with complex numbers, these are just the MacLaurin series for these functions)

By integrating the power series (1.3) we get

$$\ln(1-x) = -\sum_{k=0}^{\infty} \frac{x^{k+1}}{k+1} = -x - \frac{x^2}{2} - \frac{x^3}{3} - \cdots \qquad (|x| < 1) \qquad (1.6)$$

and, replacing x by $-x$,

$$\ln(1+x) = -\sum_{k=0}^{\infty} \frac{(-1)^{k+1} x^{k+1}}{k+1} = x - \frac{x^2}{2} + \frac{x^3}{3} - \frac{x^4}{4} \cdots \qquad (|x| < 1) \quad (1.7)$$

Don't worry if you do not remember all these details–everything you need can also be found in any Calculus text to refresh your memory. The following examples are meant to demonstrate these ideas further.

Example 1 The series in (1.7) is convergent for $x = 1$. It follows that

$$\ln 2 = 1 - \frac{1}{2} + \frac{1}{3} - \frac{1}{4} \cdots$$

Use the first 8 terms of this series to estimate $\ln 2$. How many terms would be needed to compute $\ln 2$ with an error less than 10^{-6} using this series? (Note the true value of $\ln 2 \approx 0.693\,147\,18$)

The first six terms yield

$$\ln 2 \approx 1 - \frac{1}{2} + \frac{1}{3} - \frac{1}{4} + \frac{1}{5} - \frac{1}{6} = 0.61666667$$

which has an error close to 0.08.

Since the series (1.7) is an alternating series of decreasing terms (for $0 < x \le 1$), the truncation error is smaller than the first term omitted. To force this truncation error to be smaller than 10^{-6} would therefore require that the first term omitted is smaller than $1/1,000,000$. That is, the first one million terms would suffice.

Example 2 Find the number of terms of the exponential series that are needed for $\exp x$ to have error $< 10^{-5}$ for $|x| \le 2$.

First note that the tail of the exponential series truncated after N terms increases with $|x|$. Also the truncation error for $x > 0$ will be greater than that for $-x$ since the series for $\exp(-x)$ will be alternating in sign. It is sufficient therefore to consider $x = 2$.

We shall denote by $E_N(x)$ the truncation error in the approximation using N terms:

$$\exp x \approx \sum_{k=0}^{N-1} \frac{x^k}{k!}$$

Then, we obtain, for $x > 0$

$$\begin{aligned}
E_N(x) &= \sum_{k=N}^{\infty} \frac{x^k}{k!} = \frac{x^N}{N!} + \frac{x^{N+1}}{(N+1)!} + \frac{x^{N+2}}{(N+2)!} + \cdots \\
&= \frac{x^N}{N!} \left[1 + \frac{x}{N+1} + \frac{x^2}{(N+2)(N+1)} + \cdots \right] \\
&\le \frac{x^N}{N!} \left[1 + \frac{x}{N+1} + \frac{x^2}{(N+1)^2} + \cdots \right] \\
&= \frac{x^N}{N!} \cdot \frac{1}{1 - x/(N+1)}
\end{aligned}$$

provided $x < N + 1$. For $x = 2$, this simplifies to

$$E_N(2) \le \frac{2^N}{N!} \cdot \frac{N+1}{N-1}$$

and we require this quantity to be smaller than 10^{-5}. Now $2^{11}/11! = 5.130\,671\,8 \times 10^{-5}$, while $2^{12}/12! = 8.551\,119\,7 \times 10^{-6}$. We must check the effect of the factor $\frac{N+1}{N-1} = \frac{13}{11} = 1.181\,818\,2$. Since $(1.181\,818\,2)(8.551\,119\,7) = 10.105\,869$, twelve terms is not quite sufficient. $N = 13$ terms are needed: $\frac{2^{13}}{13!} \cdot \frac{14}{12} = 1.534\,816\,4 \times 10^{-6}$.

We note that for $|x| \le 1$ in the previous example, we obtain $E_N(1) \le \frac{N+1}{N \cdot N!} < 10^{-5}$ for $N \ge 9$. For $|x| \le 1/2$, just 7 terms are needed. The number of terms required increases rapidly with x. These ideas can be used as a basis for *range reduction* so that the series would only be used for very small values of x. For example, we could use the 7 terms to obtain $e^{1/2}$ and then take

$$e^2 = \left(e^1\right)^2 = \left(e^{1/2}\right)^4$$

to obtain e^2. Greater care would be needed over the precision obtained from the series to allow for any loss of accuracy in squaring the result twice. The details are unimportant here, the point is that the series expansion can provide the basis for a good algorithm.

The magnitude of the error in any of these approximations to e^x increases rapidly with $|x|$ as can be seen from Fig. 1.3. The curve plotted is the error $e^x - \sum_{k=0}^{6} x^k/k!$. We see that the error remains very small throughout $[-1, 1]$ but rises sharply outside this interval indicating that more terms are needed there. The truncated exponential series is computed using the Python function

Fig. 1.3 Error in approximating $\exp(x)$ by a truncated series

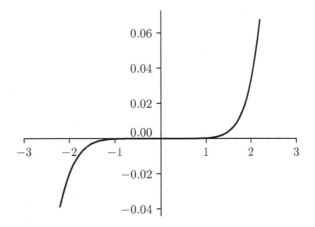

```python
import numpy as np

def expn(x, n):
    """
    Evaluates the first 'n' terms of the exponential
    series

    """

    s = np.ones_like(x)
    t = s;
    for k in range(1,n):
        t = t * x / k
        s = s + t
    return s
```

1.5 Modeling Errors Versus Errors

Computational solutions to mathematical problems are subject to a variety of different types of errors. Several of these will be discussed in greater detail in future sections. Of course, the most common, and destructive, form of error is simple human error, or making mistakes. In this book, we will live in a fantasy world where such mistakes simply do not occur! However even in this fools' paradise errors are an everyday fact of computational life.

We have already seen some simple examples of truncation error, and controlling this, in the examples above on estimating numerical values from series expansions. The focus of those examples was less on the error control and more on the need for smarter approaches to the computation to render it feasible.

The accumulated effect of rounding, or roundoff, error is another important topic in estimating the accuracy of an overall computation process. This is discussed in the next chapter in the context of the finite representation of numbers through the floating-point representation and arithmetic systems used by (almost) all computing systems. However, there is another critical source of error which usually arises much earlier in the solution process, modeling error. A simple, familiar and historical example is in the simple harmonic motion model for a simple pendulum. The model that is frequently found in both calculus and elementary physics texts is that the motion of such a simple pendulum is described by the differential equation $\theta'' = \omega^2 \sin \theta$ which is derived from the approximation $\theta \approx \sin \theta$ for small values of the angle θ (measured of course in radians). There was an excellent reason for the adoption of this model some two hundred years ago: the resulting differential equation was easily solved using paper and pencil techniques, and neither computational machinery, nor computational methods were available to solve a more accurate description of the true dynamics.

In Fig. 1.2, we see a strong emphasis on the feedback loop from the solution of our model to the original problem. If we subject this simplistic pendulum model to that examination, we see that the model provides us with a perpetual motion machine. Not only is there inherent inaccuracy in the approximation we used, but there is also neglect of any frictional or other damping effects that will certainly be present in practice. (Your grandparents' pendulum clock needs weights that can be "rewound" in order to keep going.)

The basic point here is that however good our solutions may be for the computational model we obtain, there are modeling errors as well as computational ones, and so all solutions must be examined for what they tell us about the "real-life" situation under examination–and typically they need further refinement as a result of this assessment. For many of our readers, this is a new realm of mathematics where there is no (unique, or otherwise) "right" answer. This is both the beauty and the challenge of the whole process.

1.6 Conclusions and Connections: Motivation and Background

What have you learned in this chapter? And where does it lead?

This first, short, chapter sets the stage for the remainder of the book. You have been introduced to some of the underlying issues–and even some of the ways in which these can be tackled. In many senses this chapter answers the fundamental question "Why do I need to learn this?", or even "Why do I care?"

Scientific Computing is by its very nature–almost by definition–a subject that only becomes important in its ability to help us solve real problems, and those real problems inevitably start with the need to "translate" a question from some other area of science into a mathematical model and then to solve the resulting mathematical problem.

The solution will almost always be a computational, and approximate, solution since most models are not amenable to paper and pencil approaches. Therefore we need to study the process of mathematical modeling and the fundamental topics of scientific computing per se. The latter includes the ability to code your solution algorithm. We have chosen Python as the language for this book and so this first chapter included introductions to both modeling and python programming.

Any computational solution has errors inherent in the process. There's more on that in the next chapter, but even in this introduction we see that simplistic approaches to approximating a function through a power series are often not practical, but that a careful restructuring of the problem can overcome that. This motivates the subsequent study of numerical methods and the importance of controlling errors and the practicality of any particular approach.

That is what much of the rest of the book addresses. Read on and enjoy the challenges and successes.

Exercises

1. Consider the motivating modeling problem from the beginning of this section: "There is an outbreak of the flu. Is it significant?" Brainstorm about this situation and list some assumptions, variables, and relationships. Try to come up with a model that will predict the number of infected people at a given time. You don't need to find a particular solution. What are the strengths and weaknesses of your approach? What are some challenges in seeking a solution to this problem?

2. Suppose that two stray cats, a male and a female, have a litter of kittens. Derive a sequence, using mathematical modeling, that describes how the stray cat population will grow over time. How many cats will there be in two, five, and ten years.

3. For the above cat problem, propose a humane and cost effective intervention strategy to control the population and incorporate it into your model. Compare the results.

4. Suppose that two bicyclists are traveling towards each other. Derive a mathematical model to determine when they would meet. Clearly define all the assumptions, variables, and relationships you choose to come up with the model. Create a python program to simulate this scenario and test it for a range of model parameters. Do your answers make sense? Could you explain your model to someone else clearly?

5. Write a script to approximate the natural logarithm using the first 6 terms of Eq. (1.7). Use it to estimate $\ln 1.128$.

6. How many terms of the series (1.7) are needed to approximate $\ln 1.128$ with error smaller than 10^{-4}? Evaluate this approximation and verify that the error is indeed within the tolerance.

7. Determine the number of terms of the exponential series that are needed to obtain $e^{0.1}$ with error smaller than 10^{-8}. Evaluate the sum of these terms and verify that the desired accuracy is achieved.

8. Use the series in (1.6) and (1.7) together with basic properties of the logarithm function to obtain a series expansion for $\ln\left((1+x)/(1-x)\right)$. Show that choosing $x = 1/3$ then provides a convergent series to compute $\ln 2$. Use this series to estimate $\ln 2$ with error less than 10^{-6}. Write a script to approximate the natural logarithm function using the first six terms of this series and graph the function, and its approximation, and its error for x in $(0, 3)$.

9. Write a script to estimate the natural logarithm function using n terms of a series expansion. Graph the error in this approximation to $\ln x$ for $x \in (0, 3)$ using 10 terms.

10. Estimate π using the identity $\arctan 1/\sqrt{3} = \pi/6$ by considering a series expansion $\arctan x$. How many terms are needed to have an error of less than 10^{-3}?

11. The erf function, or "error function", defined by

$$\text{erf}(x) = \frac{2}{\sqrt{\pi}} \int_0^x \exp\left(-t^2\right) dt$$

is an important function in statistics. Derive the series expansion

$$\text{erf}\,(x) = \frac{2}{\sqrt{\pi}} \sum_{k=0}^{\infty} \frac{(-1)^k \, x^{2k+1}}{(2k+1)\,k!}$$

Use the first 10 terms of this series to obtain a graph of this function over the interval $[0, 4]$.

12. A car is moving at 18 m/s with an instantaneous acceleration of 1.7 m/s^2. Build a second degree Taylor polynomial to estimate how far the car goes in the next second, two seconds, and ten seconds. Discuss at which point you think the polynomial fails to be a good approximation to estimate the distance that car travels.

13. In *Calculus Early Transcendentals* 8th Edition by James Stewart they reference *Physics: Calculus* 2nd Edition by Eugene Hecht in deriving the equations to determine the period of a pendulum of length L they get

$$T = 2\pi\sqrt{L/g}.$$

The authors obtain the equation

$$\alpha_T = -g\sin(\theta)$$

for the tangential acceleration of the bob of the pendulum and claim "for small angles, the value of θ in radians is very nearly the value of $\sin(\theta)$; they differ by less than 2% out to about 20°."

(a) Explain how $\sin(x) \approx x$ is just a first degree Taylor polynomial approximation.

(b) Write a script to determine the range of x values for which the error in that approximation is within 2%. Does Hecht's statement hold?

Number Representations and Errors

2

2.1 Introduction

The primary objective of this book is to connect scientific computing to solving real-world problems by providing the appropriate mathematical tools, skills, and insight. Computation in its simplest form relies on first representing numbers on a computer and then understanding how arithmetic operations are impacted by that representation. In this chapter we focus on that preliminary idea and then throughout the text, remind the reader why it is important. As a somewhat alarming example, we recall an attempt to intercept a Scud missile failed due to the internal computer system's inability to represent 1/10 accurately, resulting in 28 service men dying. Ultimately a round-off error of only 9.5×10^{-8} s accumulated to be large enough for the Patriot missile to entirely miss its target by roughly 0.3 s. To read more about this incident, see *Roundoff Error and the Patriot Missile* Robert Skeel, SIAM News, July 1992, Volume 25, Number 4, page 11.

Across scientific disciplines these types of errors can impact results. Within a mathematical context, they may arise in the simple evaluation of a function, solutions of linear and nonlinear equations, and the solution of differential equations. A key take-away message is that pencil-and-paper computations may not match the output from a computer code designed to do the same thing. Moreover, as problems get more realistic, pencil-and-paper approaches no longer apply and approximate solutions must be sought with care taken in how accurate that approximation is. After understanding the fundamental ideas behind number representation, we discuss sources and types of errors that can occur and how to measure them and when possible, prevent them.

© Springer International Publishing AG, part of Springer Nature 2018
P. R. Turner et al., *Applied Scientific Computing*, Texts in Computer Science,
https://doi.org/10.1007/978-3-319-89575-8_2

2.2 Floating-Point Numbers

Computational accuracy is unfortunately limited by the fact that a computer can store only a finite subset of the infinite set of real numbers. It helps to have an understanding about the range of numbers that can be stored in a computer and their accuracy. *Floating-point representation* is the way to describe how numbers are stored on a computer. You are likely already familiar with a similar idea if you have used scientific notation. While people are used to the decimal system–using base 10 (we have ten fingers!), computer systems use binary floating-point representation (base 2).

In general, a positive number x can be expressed in terms of an arbitrary base β as

$$x = a_n\beta^n + a_{n-1}\beta^{n-1} + \cdots + a_1\beta^1 + a_0\beta^0 + a_{-1}\beta^{-1} + a_{-2}\beta^{-2} + \cdots + a_{-m}\beta^{-m}$$

and each of the base-β digits $a_n, a_{n-1}, \ldots, a_1, a_0, a_{-1}, a_{-2}, \ldots, a_{-m}$ is an integer in the range $0, 1, 2, \ldots, \beta - 1$. The representation above could be rewritten using summation notation as

$$x = \sum_{k=-m}^{n} a_k\beta^k$$

The floating-point form of x can be expressed as

$$x = f \times \beta^E \tag{2.1}$$

where the mantissa f satisfies

$$1 \le f < \beta \tag{2.2}$$

This last condition (2.2) is the basis of the *normalized* floating-point system used by computers. To approximate the mantissa with only a fixed number of significant figures, f takes on the form

$$f = b_0 + b_1\beta^{-1} + b_2\beta^{-2} + \cdots + b_{-N}\beta^N = \sum_{k=0}^{N} b_k\beta^{-k}. \tag{2.3}$$

Since in the *binary* system $\beta = 2$, Eq. (2.2) reduces to $1 \le f < 2$ so that

$$b_0 = 1 \tag{2.4}$$

in the binary representation (2.3) of the mantissa. This first bit is not stored explicitly in normalized binary floating-point representations and is called the *implicit* bit. Keep in mind, that bit (although not stored) must be used in any computations.

Throughout the book, we may consider hypothetical computers that have a specified number of *significant figures* in order to illustrate certain ideas easily. Remember, significant figures are all digits after the first nonzero one that determine the magnitude of a number with several rules that apply to zero; zeros between nonzero digits are significant, when there is a decimal–trailing zeros are significant and leading zeros are not. For example, 123.45 and 0.012345 both have five significant figures while 10.2345 and 0.0123450 have six.

Example 1 Floating-point representations of π.

To demonstrate the above ideas, consider π. The constant $\pi = 3.141593$ to six decimal places, or seven significant figures, would be represented in a normalized decimal floating-point system using four significant figures by

$$\pi \approx +3.142 \times 10^0$$

With six significant figures, we would have

$$\pi \approx +3.14159 \times 10^0$$

This representation consists of three pieces: the *sign*, the *exponent* (in this case 0) and the *mantissa*, or *fraction*, 3.14 . . ..

In the *binary* floating-point system using 18 significant binary digits, or *bits* , we would have

$$\pi \approx +1.10010010000111111 \times 2^1$$

Here the digits following the *binary point* represent increasing powers of $1/2$. Thus the first five bits represent

$$\left(1 + \frac{1}{2} + \frac{0}{4} + \frac{0}{8} + \frac{1}{16}\right) \times 2^1 = 3.125$$

Note that most computer systems and programming languages allow quantities which are known to be (or declared as) integers to be represented in their exact binary form. However, this restricts the size of integers which can be stored in a fixed number of bits (referred to as a *word*). This approach does help avoid the introduction of any further errors as the result of the representation. The IEEE binary floating-point standards is the standard floating point representation used in practice (See IEEE Standard 754, Binary Floating-point Arithmetic, Institute for Electrical and Electronic Engineers, New York 2008). There are two formats that differ in the number of bits used. IEEE single precision uses a 32-bit *word* to represent the sign, exponent and mantissa and double precision uses 64-bits . Table 2.1 shows how these bits are used to store the sign, exponent, and mantissa.

In every day computations, we are used to simply rounding numbers to the nearest decimal. This is generalized by the notion of *symmetric rounding* in which a number is represented by the nearest member of the set of representable numbers (such as in Example 1). Be aware though that this may not be the approach taken by your

Table 2.1 IEEE normalized
binary floating-point
representations

Numbers of bits	Single precision	Double precision
Sign	1	1
Exponent	8	11
Mantissa (including implicit bit)	24	53

computer system. The IEEE standard requires the inclusion of symmetric rounding, *rounding towards zero* (or *chopping*), and the directed rounding modes towards either $+\infty$ or $-\infty$.

Consider Example 1 again. Using chopping in the decimal floating-point approximations of π gives $\pi \approx +3.141 \times 10^0$ and with 7 significant figures would give $\pi \approx +3.1415926 \times 10^0$. Throughout this text, symmetric rounding will be used but the reader should remember that there may be additional details to consider depending on the computer systems he or she is using.

It may be becoming clear that the errors in the floating-point representation of real numbers will also impact computations and may introduce additional errors. To gain insight into this issue, consider the following examples.

Example 2 Consider the following simple operations on a hypothetical decimal computer in which the result of every operation is rounded to four significant digits.

1. The addition

$$1.234 \times 10^0 + 1.234 \times 10^{-1} = 1.3574 \times 10^0$$
$$\approx 1.357 \times 10^0$$

has a rounding error of 4×10^{-4} which is also true of the corresponding subtraction

$$1.234 \times 10^0 - 1.234 \times 10^{-1} = 1.1106 \times 10^0$$
$$\approx 1.111 \times 10^0$$

Multiplication of the same pair of numbers has the exact result 1.522756×10^{-1} which would be rounded to 1.523×10^{-1}. Again there is a small rounding error.

2. The somewhat more complicated piece of arithmetic

$$\frac{1.234}{0.1234} - \frac{1.234}{0.1233}$$

demonstrates some of the pitfalls more dramatically. Proceeding in the order suggested by the layout of the formula, we obtain

$$\frac{1.234}{0.1234} \approx 1.000 \times 10^1, \text{ and}$$
$$\frac{1.234}{0.1233} \approx 1.001 \times 10^1$$

from which we get the result -0.001×10^1 which becomes -1.000×10^{-2} after normalization.

3. If we perform the last calculation rather differently, we can first compute

$$\frac{1}{0.1234} - \frac{1}{0.1233} \approx 8.104 - 8.110 = -6.000 \times 10^{-3}$$

Multiplying this result by 1.234 we get -7.404×10^{-3} which is much closer to the correct result which, rounded to the same precision, is -8.110×10^{-3}.

The examples above highlight that, because number representation is not exact on a computer, additional errors occur from basic arithmetic. The subtraction of two numbers of similar magnitudes can be especially troublesome. (The result in part 2 of Example 2 has only *one* significant figure since the zeros introduced after the normalization are negligible.) Although this may seem alarming, part 3 demonstrates that by being careful with the order in which a particular computation takes place, errors can be avoided.

Something else to be aware of–problems can arise from the fact that a computer can only store a finite set of numbers. Suppose that, on the same hypothetical machine as we used in Example 2 there is just one (decimal) digit for the exponent of a floating-point number. Then, since

$$\begin{aligned}
\left(3.456 \times 10^3\right) \times \left(3.456 \times 10^7\right) &= 3456 \times 34560000 \\
&= 119439360000 \\
&\approx 1.194 \times 10^{11}
\end{aligned}$$

the result of this operation is too large to be represented in our hypothetical machine. This is called *overflow*. Similarly, if the (absolute value of a) result is too small to be represented, it is called *underflow*. For our hypothetical computer this happens when the exponent of the normalized result is less than -9.

These ideas indicate that even some seemingly simple computations must be programmed carefully in order to avoid overflow or underflow, which may cause a program to fail (although many systems will not fail on underflow but will simply set such results to zero, but that can often result in a later failure, or in meaningless answers). The take-away message is not to stress too much about these issues, but be aware of them when solutions seem questionable or a program fails to execute.

2.2.1 Python Number Representation

Most computing platforms use only one type of number – IEEE double precision floating-point. Dictated by the computer's hardware, Python is no exception to this. On typical computers, Python floating-point numbers map to the IEEE double precision type. Integers in Python are represented separately as integer data types, so there is no need to worry about numerical precision as long as operations only involve integers.

From Table 2.1, we see that the double-precision representation uses 11 bits for the binary exponent which therefore ranges from about -2^{10} to $2^{10} = 1024$. (The actual range is not exactly this because of special representations for small numbers and for $\pm\infty$.) The mantissa has 53 bits including the implicit bit. If $x = f \times 2^E$ is a normalized floating-point number then $f \in [1, 2)$ is represented by

$$ f = 1 + \sum_{k=1}^{52} b_k 2^{-k} $$

Since $2^{10} = 1024 \approx 10^3$, these 53 significant bits are equivalent to approximately 16 significant decimal digits accuracy.

The fact that the mantissa has 52 bits *after the binary point* means that the next machine number greater than 1 is $1 + 2^{-52}$. This gap is called the *machine unit*, or machine *epsilon*. This and other key constants of Python's arithmetic are easily obtained from the sys module (import sys first).

Python variable	Meaning	Value
sys.float_info.epsilon	Machine unit	$2^{-52} \approx 2.220446049250313\mathrm{e}{-16}$
sys.float_info.min	Smallest positive number	$2.2250738585072014\mathrm{e}{-308}$
sys.float_info.max	Largest positive number	$1.7976931348623157\mathrm{e}{+308}$

Here, and throughout, the notation $x\mathrm{e}n$ where x, n are numbers stands for $x \times 10^n$. The quantity 10^{-8} may be written in practice as $1\mathrm{e}{-8}$.

In most cases we wish to use the NumPy package for Python for numerical computations. NumPy defines its own floating-point and integer data types offering choice between different levels of precision, e.g. numpy.float16, numpy.float32, and numpy.float64. These types are compatible with the built-in floating-point type of ordinary Python and numpy.float64 (double precision) is the default type. NumPy offers access to an extended precision 80-bit floating-point types as well on operating systems where this is available. Somewhat confusingly, this data type is available as numpy.float96 and numpy.float128. The latter two types might seem to suggest that they offer 96-bit and 128-bit precision, respectively, but the name only relates to how many bits the variables are zero-padded to in memory. They still offer only 80 bits' precision at most, and only 64 bits by default on Windows. We advise only using NumPy's default numpy.float64 type unless special requirements dictate otherwise and you know your platform actually offers the higher precision.

The machine precision-related constants of NumPy's arithmetic can be obtained using NumPy's finfo function (import numpy first).

NumPy function call	Meaning	Value
numpy.finfo(numpy.float64).eps	Machine unit	$2^{-52} \approx 2.2204460492503131e{-}16$
numpy.finfo(numpy.float64).tiny	Smallest positive number	2.2250738585072014e−308
numpy.finfo(numpy.float64).max	Largest positive number	1.7976931348623157e+308

In Python, neither underflow nor overflow cause a "program" to stop. Underflow is replaced by zero, while overflow is replaced by $\pm\infty$. This allows subsequent instructions to be executed and may permit meaningful results. Frequently, however, it will result in meaningless answers such as $\pm\infty$ or NaN, which stands for *Not-a-Number*. NaN is the result of indeterminate arithmetic operations such as $0/0$, ∞/∞, $0 \cdot \infty$, $\infty - \infty$ etc.

2.3 Sources of Errors

In this section, we consider three primary sources of error, although the topic of error analysis and error control is in and of itself worthy of an entire text. The purpose here is to introduce the reader to these ideas so later we can use them to understand the behavior of numerical methods and the quality of the resulting solutions.

2.3.1 Rounding Errors

We have already seen that these arise from the storage of numbers to a fixed number of binary or decimal places or significant figures.

Example 3 The equation

$$x^2 - 50.2x + 10 = 0$$

factors into $(x - 50)(x - 0.2) = 0$, thus the roots are 50 and 0.2. Approaching this problem using an algorithm could be done with the quadratic formula. Consider using a hypothetical four decimal digit machine with the quadratic formula. The roots obtained are 50.00 and 0.2002. The errors are due to rounding of all numbers in the calculation to four significant figures.

If however, we use this same formula for the larger root together with the fact that the *product* of the roots is 10, then, of course, we obtain the values 50.00 and $10.00/50.00 = 0.2000$. Again, note that rounding errors can be addressed by the careful choice of numerical process, or the *design of the algorithm*. This approach to solving for roots of a quadratic equation is the best approach!

The previous computations in Example 3 used symmetric rounding but if chopping were used instead, the larger root would be obtained as 49.99 and the smaller would then be $10/49.99 = 0.200$ as expected.

2.3.2 Truncation Error

Truncation error is the name given to the errors which result from the many ways in which a numerical process can be cut short or truncated. Many numerical methods are based on truncated Taylor series approximations to functions and thus, this type of error is of particular importance and will be revisited throughout the text.

To get more comfortable with the notion of truncation error, consider the geometric series,

$$\frac{1}{2} + \frac{1}{4} + \frac{1}{8} \cdots = \sum_{n=1}^{\infty} \left(\frac{1}{2}\right)^n$$

If we truncate this series after only the first three terms, then we get 0.875, which gives a truncation error of 0.125.

Of course, including more terms in the sum before truncating would certainly lead to a more accurate approximation. The following examples show how to use the notion of truncation error to determine a desired accuracy.

Example 4 (a) Determine the number of terms of the standard power series for $\sin(x)$ *needed to compute* $\sin(1/2)$ *to five significant figures.*

The series

$$\sin(x) = x - \frac{x^3}{3!} + \frac{x^5}{5!} - \cdots = \sum_{n=0}^{\infty} (-1)^n \frac{x^{2n+1}}{(2n+1)!}$$

has infinite radius of convergence, so we need not worry about the issue of convergence for $x = 1/2$. Indeed the terms of the series in this case have decreasing magnitude and alternating sign. Therefore the error in truncating the series after N terms is bounded by the first term omitted. We have already seen that $\sin(x)$ can be well approximated by x itself for small values so the true value is expected to be a little less than 0.5. Accuracy to five significant figures is therefore achieved if the absolute error is smaller than 5×10^{-6}. Thus we seek N such that

$$\frac{1}{2^{2N+1}(2N+1)!} < 5 \times 10^{-6},$$

or equivalently

$$2^{2N+1}(2N+1)! > 200,000.$$

The first few odd factorials are 1, 6, 120, 5040, and the first few odd powers of 2 are 2, 8, 64, 512. Now $64 \times 120 = 7,680$ is too small, while $512 \times 5040 = 2,580,480$ readily exceeds our threshold. Therefore the first three terms suffice, so that to five significant figures we have

$$sin(1/2) = \frac{1}{2} - 18 \times 6 + 164 \times 120 = 0.47917.$$

(b) Determine the number of terms N of the exponential series required to estimate
$\sqrt{e}$ *to five decimal places.*

This time we need to use the exponential series to compute $e^{1/2}$. Again, convergence of the power series

$$e = \exp(1) = 1 + 1 + \frac{1}{2!} + \frac{1}{3!} + \cdots = \sum_{n=0}^{\infty} \frac{1}{n!}$$

isn't an issue since it has an infinite radius of convergence.

The terms of the series decrease rapidly. But this time it is not sufficient to merely find the first term smaller than the required precision since the tail of the series could conceivably accumulate to a value greater than that tolerance. What is that tolerance in this case? We know that $e > 1$, and so therefore is its square root. In this case then, five significant figures equates to four decimal places. That is we need the error to be bounded by 5×10^{-5}.

Truncating the exponential series for $x = 1/2$ after N terms means the omitted tail of the series is

$$\sum_{n=N}^{\infty} \frac{1}{2^n n!} = \frac{1}{2^N N!} + \frac{1}{2^{N+1}(N+1)!} + \frac{1}{2^{N+2}(N+2)!} + \cdots$$

Since each successive factorial in the denominator increases by a factor of at least (N+1), we can bound this tail by the geometric series

$$\frac{1}{2^N N!}\left(1 + \frac{1}{2(N+1)} + \frac{1}{[2(N+1)]^2} + \cdots\right).$$

The sum of this geometric series is

$$\frac{1}{1 - \frac{1}{2N+2}} = \frac{2N+2}{2N+1}$$

so that the complete tail is bounded by

$$\frac{1}{2^N N!}\frac{2N+2}{2N+1}.$$

The first few values for $N = 0, 1, 2, , 7$ are $2, 0.666667, 0.15, 0.02381, 0.002894$, $0.000284, 2.34 \times 10^{-05}, 1.65 \times 10^{-06}$. We see that the required tolerance is achieved by 2.34×10^{-5} for $N = 6$.

2.3.3 Ill-Conditioning

Given that the underlying model is sound and the numerical method is "stable" (more on that later), there still may be inherent issues in the underlying problem you are trying to solve. This notion is called the conditioning of the problem and is based on the idea that small changes to the input should lead to small changes in the output. When that is not the case, we say the underlying problem is *ill-conditioned*.

A classical example of ill-conditioning is the so-called Wilkinson polynomial. Consider finding the roots of

$$p(x) = (x - 1)(x - 2) \cdots (x - 20) = 0$$

where p is given in the form

$$p(x) = x^{20} + a_{19}x^{19} + \cdots + a_1 x + a_0$$

If the coefficient $a_{19} = -210$ is changed by about 2^{-22}, then the resulting polynomial has only 10 real zeros, and five complex conjugate pairs. The largest real root is now at $x \approx 20.85$. A change of only one part in about 2 billion in just one coefficient has certainly had a significant effect!

Underlying ill-conditioning is often less of an issue than modeling errors or for example, using an inappropriate method to solve a specific problem. However, understanding the behavior of the underlying problem (or model or method) you have in front of you is always important so that you can make the right choices moving forward. This idea will be revisited throughout the text.

2.4   Measures of Error and Precision

The purpose of numerical methods in general is to approximate the solution to a problem, so the notion of accuracy and errors is critical. Suppose that $\widehat{x}$ is an approximation to a number x. The *absolute error* is defined to be $|x - \widehat{x}|$; this corresponds to the idea that x and $\widehat{x}$ agree to a number of decimal places. However, if I tell you the absolute error is 2, is that always valuable information? If you were originally trying to approximate the number 4 with the number 2, maybe not. If you were trying to approximate 100,000 with 100,002, then maybe so.

The *relative error* is usually given by $|x - \widehat{x}| / |x| = |1 - \widehat{x}/x|$ which corresponds to agreement to a number of significant figures. In the case that we are trying to approximate x with $\widehat{x}$, using a numerical approach then we wouldn't even know x in advance. In practice, the x in the denominator can be replaced by its approximation $\widehat{x}$ and the best one can do is to try to bound errors to obtain a desired accuracy.

In this case the relative error is given by $|x - \widehat{x}| / |\widehat{x}| = |1 - x/\widehat{x}|$. (The asymmetric nature of this definition is somewhat unsatisfactory, but this can be overcome by the alternative notion of relative precision.)

Example 5 Find the absolute and relative errors in approximating e by 2.7183. What are the corresponding errors in the approximation $100e \approx 271.83$?

Unfortunately, we cannot find the absolute errors exactly since we do not know a numerical value for e *exactly*. However we can get an idea of these errors by using the more accurate approximation

$$e \approx 2.718281828459046.$$

Then the absolute error can be estimated as

$$|e - \widehat{e}| = |2.718281828459046 - 2.7183| = 1.8172 \times 10^{-5}$$

The corresponding relative error is

$$\frac{|e - \widehat{e}|}{|e|} = \frac{|2.718281828459046 - 2.7183|}{2.718281828459046} = 2.6.6849 \times 10^{-6}$$

Now using the approximations to $100e$, we get the absolute error

$$|271.8281828459046 - 271.83| = 1.8172 \times 10^{-3}$$

which is exactly 100 times the error in approximating e, of course. The relative error is however unchanged since the factor of 100 affects numerator and denominator the same way:

$$\frac{|271.8281828459046 - 271.83|}{271.8281828459046} = \frac{|2.718281828459046 - 2.7183|}{2.718281828459046} = 2.6.6849 \times 10^{-6}$$

As a general principle, we expect absolute error to be more appropriate for quantities close to one, while relative error seems more natural for large numbers or those close to zero. A relative error of 5 or 10% in computing and/or manufacturing a bridge span has serious potential–either a "Bridge to Nowhere" or "A Bridge Too Far" being possible outcomes. This is a situation where absolute error must be controlled even though it may represent only a very small relative error. On the other hand a small absolute error in computing the Dow Jones Industrial Average (currently at the time of this publication, around 25,000) is not important whereas a significant relative error could cause chaos in financial markets around the world. To put that in perspective a 5% drop would almost certainly end up with that day of the week being labeled black. This is certainly a situation where relative error is the only one that really matters and it must be controlled to a very low level.

Although often we want to know the error in a particular number, there are times we actually want to approximate a function with another function. In our discussion of the pendulum, $\sin \theta$ is approximated with θ. Throughout the section on series expansions, the underlying idea is to indeed approximate a complicated function with a simpler polynomial. We need different metrics in order to be able to even define what that error means. The three most commonly used metrics are defined for approximating a function f on an interval $[a, b]$ by another function (for example, a polynomial) p:

The *supremum* or *uniform* or L_∞ *metric*

$$||f - p||_\infty = \max_{a \le x \le b} |f(x) - p(x)|.$$

The L_1 *metric*

$$||f - p||_1 = \int_a^b |f(x) - p(x)| \, dx.$$

The L_2 or *least-squares metric*

$$||f - p||_2 = \sqrt{\int_a^b |f(x) - p(x)|^2 \, dx}.$$

These metrics are often referred to as *norms*. The first measures the extreme discrepancy between f and the approximating function p while the others are both measures of the "total discrepancy" over the interval. The following examples provide insight.

Example 6 Find the L_∞, L_1, and L_2 errors in approximating $\sin \theta$ by θ over the interval $[0, \pi/4]$

The plot of the two functions in Fig. 2.1 helps shed light on what we are measuring. The L_∞ error requires the maximum discrepancy between the two curves. From the graphs, or using calculus, one can deduce that occurs at $\theta = \pi/4$ so that

$$||\sin \theta - \theta||_\infty = \pi/4 - \sin(\pi/4) = 7.829\,1 \times 10^{-2}$$

Fig. 2.1 Error in approximating $\sin(\theta)$ by θ

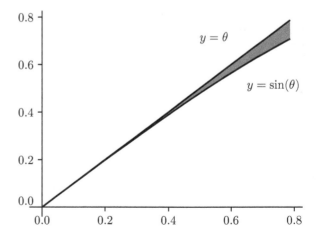

The L_1 error is the area of the shaded region between the curves:

$$||\sin\theta - \theta||_1 = \int_0^{\pi/4} (\theta - \sin\theta)\, d\theta = \frac{1}{32}\pi^2 + \frac{1}{2}\sqrt{2} - 1 = 1.5532 \times 10^{-2}$$

The L_2 error is given by

$$||\sin\theta - \theta||_2^2 = \int_0^{\pi/4} (\theta - \sin\theta)^2\, d\theta = 6.9728 \times 10^{-4}$$

so that $||\sin\theta - \theta||_2 = \sqrt{6.9728 \times 10^{-4}} = 2.6406 \times 10^{-2}$.

In some cases we only know values of f at a discrete set of points $x_0, x_1, \ldots, x_N$. The corresponding discrete measures are given by:

$$\max_{k=0,1,\ldots,N} |f(x_k) - p(x_k)|$$

$$\sum_{k=0}^{N} |f(x_k) - p(x_k)|, \quad \text{and}$$

$$\sqrt{\sum_{k=0}^{N} |f(x_k) - p(x_k)|^2}$$

The last measure arises when trying fit to experimental data to a model and is called, a *least-squares* measure. All of these are still referred to as norms as well.

2.5   Floating-Point Arithmetic

With the previous ideas for measuring and interpreting errors, we briefly revisit floating-point arithmetic and the importance of algorithm design. In what follows, consider an arbitrary base β but keeping in mind this is 2 for computers and 10 for calculators.

Recall that a positive number x will be represented in normalized floating-point form by

$$\hat{x} = \hat{f}\beta^E \tag{2.5}$$

where the fraction $\hat{f}$ consists of a fixed number, $(N+1)$ say, of base-β digits:

$$\hat{f} = \widehat{d_0}.\widehat{d_1}\widehat{d_2}\ldots\widehat{d_N} \tag{2.6}$$

where $1 \le \widehat{d_0} < \beta$ and $0 \le \widehat{d_k} < \beta$ for $k = 1, 2, \ldots, N$.

The absolute error in representing $x = f\beta^E = \beta^E(d_0.d_1d_2\ldots)$ by $\hat{x}$ is therefore

$$|x - \hat{x}| = \beta^E |f - \hat{f}| \tag{2.7}$$

If $\widehat{f}$ is obtained from f by chopping then $\widehat{d_k} = d_k$ for $k = 0, 1, \ldots, N$ so that (2.7) becomes

$$|x - \widehat{x}| = \beta^E \sum_{k=N+1}^{\infty} d_k \beta^{-k} \leq \beta^{E-N}$$

Symmetric rounding is equivalent to first adding $\beta/2$ to d_{N+1} and then chopping the result. It follows that, for symmetric rounding,

$$|x - \widehat{x}| \leq \frac{\beta^{E-N}}{2} \tag{2.8}$$

Of more interest for floating-point computation is the size of the relative error:

$$\frac{|x - \widehat{x}|}{|x|} = \frac{|f - \widehat{f}|}{|f|}$$

In the same manner as above, we find that

$$\frac{|f - \widehat{f}|}{|f|} \leq \begin{cases} \frac{1}{\beta^N} & \text{for chopping} \\ \frac{1}{2\beta^N} & \text{for symmetric rounding} \end{cases} \tag{2.9}$$

To prove the "chopping" result, note that

$$\frac{|f - \widehat{f}|}{|f|} = \frac{\sum_{k=N+1}^{\infty} d_k \beta^{-k}}{|f|} \leq \frac{\beta^{-N}}{|f|} \leq \beta^{-N}$$

since $|f| \geq 1$ and $d_k \leq \beta - 1$ for each k. The "rounding" result is proved similarly.

Next, let's consider measuring how the errors arising from floating point representation can propagate through computations. One of the requirements of the IEEE floating-point standards is that the stored result of any floating-point operation equals the correctly rounded value of the *exact* result assuming the data were themselves exact.

This requirement implies that the final rounding and normalization result in a relative error no greater than the bounds in (2.9). However this is not the only source of error since the data are usually not exact. If $\blacklozenge$ in $x \blacklozenge y$ stands for one of the standard arithmetic operations $+, -, \times, \div$ then to analyze floating-point arithmetic errors fully we require bounds on the quantity

$$\frac{\left|\widehat{(x \blacklozenge y)} - (x \blacklozenge y)\right|}{|(x \blacklozenge y)|}$$

Such bounds vary according to the nature of the operation $\blacklozenge$. Note that this level of analysis is not common in practice for many real-world applications, but the reader should be aware that indeed thought (and theory) has been given to these ideas!

The following examples will provide further insight into the errors associated with floating-point arithmetic.

Example 7 Subtraction of nearly equal numbers can result in large relative rounding errors.

Let $x = 0.12344$, $y = 0.12351$ and suppose we compute the difference $x - y$ using a four decimal digit floating-point computer. Then

$$\widehat{x} = (1.234)\, 10^{-1}, \ \widehat{y} = (1.235)\, 10^{-1}$$

so that

$$\widehat{(x - y)} = \widehat{x} - \widehat{y} = -(0.001)\, 10^{-1} = -(1.000)\, 10^{-4}$$

while $x - y = -(0.7)\, 10^{-4}$. The relative error is therefore

$$\frac{(1.0 - 0.7)\, 10^{-4}}{(0.7)\, 10^{-4}} = \frac{3}{7} = 0.428\,57$$

The error is approximately 43%.

The next example, in terms of absolute errors, illustrates a common drawback of simplistic error bounds: they are often unrealistically pessimistic.

Example 8 Using "exact" arithmetic

$$\frac{1}{3} + \frac{1}{4} + \cdots + \frac{1}{10} = 1.428968$$

to six decimal places. However, rounding each term to four decimal places we obtain the approximate sum 1.4290.

The accumulated rounding error is only about 3×10^{-5}. The error for each term is bounded by 5×10^{-5}, so that the cumulative error bound for the sum is 4×10^{-4} which is an order of magnitude bigger the actual error committed, demonstrating that error bounds can certainly be significantly worse than the actual error.

This phenomenon is by no means unusual and leads to the study of probabilistic error analyses for floating-point calculations. For such analyses to be reliable, it is important to have a good model for the distribution of numbers as they arise in scientific computing. It is a well-established fact that the fractions of floating-point numbers are logarithmically distributed. One immediate implication of this distribution is the rather surprising statement that the proportion of (base-β) floating-point numbers with leading digit n is given by

$$\log_\beta \left(\frac{n + 1}{n} \right) = \frac{\log(n + 1) - \log n}{\log \beta}$$

In particular, 30% of decimal numbers have leading significant digit 1 while only about 4.6% have leading digit 9.

Example 9 The base of your computer's arithmetic is 2

Let n be any positive integer and let $x = 1/n$. Convince yourself that $(n + 1) x - 1 = x$. The Python command

```
x = (n + 1) * x - 1
```

should leave the value of x unchanged no matter how many times this is repeated. The table below shows the effect of doing this ten and thirty times for each $n = 1, 2, \ldots, 10$. The code used to generate one of these columns was:

```
>>> a = []
    for n in range(1,11):
        x = 1/n
        for k in range(10):
            x = (n + 1) * x - 1
        a.append(x)
```

n	$x = 1/n$	$k = 0, \ldots, 9$	$k = 0, \ldots, 29$
1	1.00000000000000	1.00000000000000	1
2	0.50000000000000	0.50000000000000	0.5
3	0.33333333333333	0.33333333331393	−21
4	0.25000000000000	0.25000000000000	0.25
5	0.20000000000000	0.20000000179016	6.5451e+06
6	0.16666666666667	0.16666666069313	−4.76642e+08
7	0.14285714285714	0.14285713434219	−9.81707e+09
8	0.12500000000000	0.12500000000000	0.125
9	0.11111111111111	0.11111116045436	4.93432e+12
10	0.10000000000000	0.10000020942783	1.40893e+14

Initially the results of ten iterations appear to conform to the expected results, that x remains unchanged. However the third column shows that several of these values are already contaminated with errors. By the time thirty iterations are performed, many of the answers are not recognizable as being $1/n$.

On closer inspection, note that the ones which remain fixed are those where n is a power of 2. This makes some sense since such numbers are representable *exactly* in the binary floating-point system. For all the others there is some initially small representation error: it is the propagation of this error through this computation which results in final (large!) error.

For example with $n = 3$, the initial value $\widehat{x}$ is not exactly 1/3 but is instead

$$\widehat{x} = \frac{1}{3} + \delta$$

where δ is the rounding error in the floating-point representation of 1/3. Now the operation

```
x = (n + 1) * x - 1
```

has the effect of magnifying this error. One iteration, gives (ignoring the final rounding error)

$$x = 4\widehat{x} - 1 = 4\left(\frac{1}{3} + \delta\right) - 1 = \frac{1}{3} + 4\delta$$

which shows that the error is increasing by a factor of 4 with each iteration.

In the particular case of $1/3$, it can be shown that the initial representation error in IEEE double precision floating-point arithmetic is approximately $-\frac{1}{3}2^{-54}$. Multiplying this by $4^{30} = 2^{60}$ yields an estimate of the final error equal to $-\frac{1}{3}2^{6} = -21.333$ which explains the entry -21 in the final column for $n = 3$.

Try repeating this experiment to verify that your calculator uses base 10 for its arithmetic.

2.6   Conclusions and Connections: Number Representation and Errors

What have you learned in this chapter? And where does it lead?

This chapter focused on the types of errors inherent in numerical processes. The most basic source of these errors arises from the fact that arithmetic in the computer is not exact except in very special circumstances. Even the representation of numbers–the floating point system–has errors inherent in it because numbers are represented in finite binary "words".

The binary system is almost universal but even in systems using other bases the same basic issues are present. Errors in representation lead inexorably to errors in arithmetic and evaluation of the elementary functions that are so critical to almost all mathematical models and their solutions. Much of this impacts rounding errors which typically accumulate during a lengthy computation and so must be controlled as much as possible. The IEEE Floating Point Standards help us know what bounds there are on these rounding errors, and therefore to estimate them, or even mitigate them.

It is not only roundoff error that is impacted by the finite nature of the computer. The numerical processes are themselves finite–unless we are infinitely patient. The finiteness of processes gives rise to truncation errors. At the simplest level this might be just restricting the number of terms in a series that we compute. In other settings it might be a spatial, or temporal, discrete "step-size" that is used. We'll meet more situations as we proceed.

Remember that truncation and rounding errors interact. We cannot control one independent of the other. Theoretically we might get a better series approximation to a function by taking more terms and so controlling the truncation error. However trying to do this will often prove futile because rounding errors will render those additional terms ineffective.

Take all this together with the fact that poor models are inevitably only approximate representations of reality, and you could be forgiven for thinking we are embarking on a mission that is doomed. That is not the case as we'll soon see. That is what much of the rest of the book addresses. Read on and enjoy the challenges and successes, starting in the next chapter with numerical approaches to fundamental calculus computations.

Exercises

1. Express the base of natural logarithms e as a normalized floating-point number, using both chopping and symmetric rounding, for each of the following systems

   (a) base 10 with three significant figures
   (b) base 10 with four significant figures
   (c) base 2 with 10 significant bits

2. Write down the normalized binary floating-point representations of 1/3, 1/5, 1/6, 1/7, 1/9, and 1/10, Use enough bits in the mantissa to see the recurring patterns.
3. Perform the calculations of Example 3. Repeat the same computation using chopping on a hypothetical 5 decimal digit machine.
4. Write a computer program to "sum" the geometric series $1 + \frac{1}{2} + \frac{1}{4} + \cdots$ stopping when

   (a) all subsequent terms are zero to four decimal places, and
   (b) two successive "sums" are equal to five decimal places.

5. How many terms of the series expansion

$$\cosh x = 1 + \frac{x^2}{2!} + \frac{x^4}{41} + \cdots = \sum_{k=0}^{\infty} \frac{x^{2k}}{(2k)!}$$

   are needed to estimate $\cosh(1/2)$ with a truncation error less than 10^{-8}? Check that your answer achieves the desired accuracy.
6. What is the range of values of x so that the truncation error in the approximation

$$\exp x = e^x \approx \sum_{k=0}^{5} \frac{x^k}{k!}$$

   be bounded by 10^{-8}?
7. Find the absolute and relative errors in the decimal approximations of e found in problem 1 above.

8. What is the absolute error in approximating $1/6$ by 0.1667? What is the corresponding relative error?

9. Find the absolute and relative errors in approximating $1/3$, $1/5$, $1/6$, $1/7$, $1/9$, and $1/10$ by their binary floating-point representations using 12-bit mantissas.

10. Suppose the function e^x is approximated by $1 + x + \frac{x^2}{2}$ on $[0, 1]$. Find the L_∞, L_1, and L_2 measures of the error in this approximation.

11. Let $x = 1.2576$, $y = 1.2574$. For a hypothetical four decimal digit machine, write down the representations $\widehat{x}, \widehat{y}$ of x, y and find the relative errors in the stored results of $x + y$, $x - y$, xy, and x/y using

    (a) chopping, and
    (b) symmetric rounding.

12. Try to convince yourself of the validity of the statement that floating-point numbers are logarithmically distributed using the following experiment.

    (a) Write computer code which finds the leading significant (decimal) digit of a number in $[0, 1)$. (Hint: keep multiplying the number by 10 until it is in $[1, 10)$ and then use the **floor** function to find its integer part.)
    (b) Use a random number generator to obtain vectors of uniformly distributed numbers in $[0, 1)$. Form 1000 products of pairs of such numbers and find how many of these have leading significant digit $1, 2, \ldots, 9$.
    (c) Repeat this with products of three, four and five such factors.

13. Repeat the previous exercise but find the leading *hexadecimal* (base 16) digits. See how closely these conform to the logarithmic distribution.

14. Consider Example 9. Try to explain the entry in the final column for $n = 5$. Use an argument similar to that used for $n = 3$. (You will find it easier if you just *bound* the initial representation error rather than trying to estimate it accurately.)

Numerical Calculus

3

3.1 Introduction

With the notion of how numbers are stored on a computer and how that representation impacts computation, we are ready to take the plunge into numerical calculus. At the heart of calculus I and II students learn about the theory and applications of differentiation and integration. In this chapter, many of the fundamentals tools from Calculus are revisited with a new objective: approximating solutions instead of deriving exact solutions. Many of the ideas here will be familiar to you if you look back at how you were introduced to Calculus, which makes this a somewhat gentle transition into numerical methods. But first, you may be wondering why we need such computational tools after spending several semesters learning pencil and paper Calculus. We provide some motivating examples below.

The world around us is often described using physics, specifically *conservation laws*. For example, conservation of mass, energy, electrical charge, and linear momentum. These laws state that the specific physical property doesn't change as the system evolves. Mathematically, these conservation laws often lead to partial differential equation models that are used to build large-scale simulation tools. Such tools provide insight to weather behavior, biological processes in the body, groundwater flow patterns, and more. The purpose of the underlying algorithms which comprise those simulation tools is to approximate a solution to the system of partial differential equations that models the scenario of interest. That can be achieved by using *numerical derivatives* in those partial differential equations while paying special attention to the accuracy of the approach.

Throughout this text, we will use the so-called diffusion equation as an application to demonstrate the usefulness of a variety of numerical methods. We introduce the diffusion equation briefly in the next model in the context of heat flow, but in general the diffusion equation is used in a variety of disciplines (but the coefficients have different meanings). Do not worry if you are unfamiliar with partial differential

© Springer International Publishing AG, part of Springer Nature 2018

P. R. Turner et al., *Applied Scientific Computing*, Texts in Computer Science,

https://doi.org/10.1007/978-3-319-89575-8_3

equations in general. A deep understanding will not be needed to approach the models we will be considering here. In particular, the basic notion of the diffusion equation is that the quantity of interest flows from regions of high concentration to low. For example, heat flows from hot spots to cold, or air flows from regions of high pressure to low.

Example 1 The need for numerical calculus: Heat flow on a thin insulated wire.

Consider a wire of length L whose temperature is kept at $0\,°C$ at both ends. Let $u(x, t)$ be the temperature on the wire at time t and position x with an initial temperature at $t = 0$ given as $u_0(x) = u(x, 0)$. A partial differential equation that models the temperature distribution along the wire is given by

$$\rho c \frac{\partial u}{\partial t} - K \frac{\partial^2 u}{\partial x^2} = f(x, t) \tag{3.1}$$

or

$$\rho c u_t - K u_{xx} = f(x, t).$$

Here ρ is the density, c is the specific heat, and K is the thermal conductivity of the wire. The function $f(x, t)$ is a heat source, for example from the electrical resistance of the wire.

The model is based on the assumption that the rate of change of u (i.e. u_t) is proportional to the curvature of u (i.e. u_{xx}). Thus, the sharper the curvature, the faster the rate of change. If u is linear in space then its second derivative is zero at a given point, and we say u has reached steady-state.

Suppose we wanted to approximate values of the temperature along the wire at discrete locations and at specific times until some final time, T_f. The idea is to partition the time interval $0 \le t \le T_f$ using increments of Δt and the length of the wire $0 \le x \le L$ using increments of Δx. This generates a set of discrete points in space and time, as in Fig. 3.1. We now seek $u_i^k \approx u(x_i, t^k)$, where we use t^k here to denote tk as an approximation to the temperature at those points for say $i = 0, 1, 2, \ldots n$ and $k = 0, 1, 2, \ldots m$. The next step (which we will get to later in this chapter) will be to discretize Eq. (3.1) using numerical derivatives.

Numerical differentiation also becomes necessary in circumstances where the actual derivative is not available. In some modeling situations, a function is only available from a table of values, perhaps from some experimental data, or from a call to an off-the-shelf executable (such as a simulation tool). However it may be necessary to estimate this derivative based on only those data points or simulation output values. Another example we will consider later arises in shooting methods for solving boundary value problems.

Even for functions of a single variable, the standard calculus technique for optimization of testing for critical points is not always sufficient. It is easy to find examples where setting the derivative to zero results in an equation which we cannot solve algebraically.

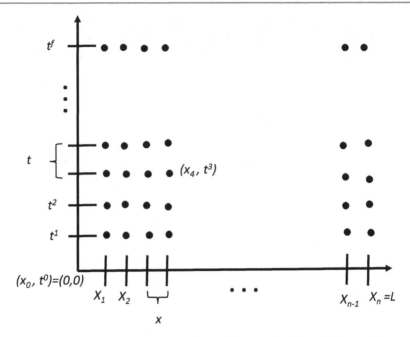

Fig. 3.1 The grid of points in space (horizontally) and time (vertically) at which we seek to approximate the temperature along the wire

When it comes to integration there is similar motivation. Some times, you may only have data or function values rather than an analytic function to integrate. Alternatively, the function may not have an analytic anti-derivative.

Example 2 The need for numerical integration:
Consider the evaluation of the standard normal distribution function of statistics

$$N_{0,1}(x) = \frac{1}{\sqrt{2\pi}} \int_{-\infty}^{x} \exp\left(-t^2/2\right) dt = \frac{1}{2} + \frac{1}{\sqrt{2\pi}} \int_{0}^{x} \exp\left(-t^2/2\right) dt \quad (3.2)$$

This integral is essential in statistical analysis, but it *cannot* be evaluated by conventional calculus techniques. Surprisingly, there is no anti-derivative for $e^{-t^2/2}$ expressible in terms of elementary functions. Instead, a function is defined to be similar to the antiderivative, called erf (but that really just gives it a name; it does not help with its evaluation). To evaluate this function, approximation methods are needed. One possibility is to use numerical integration methods similar to those discussed in later on. There are other possibilities based on series expansions which can be used for this particular integral. More on this important example later.

We should note, in practice we do not typically confront the situation where a function is known by a formula and yet we cannot differentiate it. The rules of differentiation clearly apply when a function is composed of more elementary functions.

This is the primary reason why symbolic differentiation is straightforward, while symbolic integration is not. The situation is reversed for numerical calculus in that efficient and accurate numerical integration is much easier than differentiation.

3.2 Numerical Differentiation

Think back to when you were in Calculus and first learned about the derivative (before you learned all the handy differentiation rules). Likely it was this definition:

$$f'(x_0) = \lim_{h \to 0} \frac{f(x_0 + h) - f(x_0)}{h}.$$

This was motivated by the idea that the slope of the tangent line at a point on a curve is the result from looking at slopes of secant lines between two points on the curve and letting them get closer and closer together.

That definition gives a simple approximation to the derivative with

$$f'(x_0) \approx \frac{f(x_0 + h) - f(x_0)}{h} \tag{3.3}$$

for some small h. Again, this can also be interpreted as, "the slope of the tangent line can be *approximated* by the slope of a secant line between nearby points" often referred to as a *forward finite difference approximation*. But, what is the error in this approximation and how small should h be?

Taylor's theorem can provide answers to those questions. Suppose that f'' exists, then Taylor's theorem gives

$$f(x_0 + h) = f(x_0) + hf'(x_0) + \frac{h^2}{2}f''(\xi)$$

for some ξ between x_0 and $x_0 + h$. We can now express the error in the approximation as

$$f'(x_0) - \frac{f(x_0 + h) - f(x_0)}{h} = -\frac{h}{2}f''(\xi) \tag{3.4}$$

Recall this sort of error is called the truncation error. Here, the truncation error in approximating the derivative in the way is proportional to the steplength h used in its computation. We often say that the truncation error is *first order* since h is raised to the first power. This is typically denoted as $\mathcal{O}(h)$.

Having an expression for the truncation error can be extremely valuable when it comes to implementation, especially if this approximation is embedded in a more complex computer code. The truncation error can be used to validate that your numerical method is actually coded up correctly (that is, bug-free) by conducting a grid refinement study to verify the theory holds. To see how, suppose that we approximate the derivative of a function that we know the analytic derivative of using a specific

h. We could then calculate the exact error E_h. The idea is to then start over with $h/2$ and calculate a new error $E_{h/2}$. Since our error is first order (i.e. proportional to the steplength) we must have that

$$\frac{E_h}{E_{h/2}} \approx \frac{h}{h/2} = 2.$$

A grid refinement study is the process of computing a sequence of approximations while changing h in a specific way and looking at the ratio of successive errors to ensure they are approaching the right value. In this case, if we kept halving h to generate more and more approximations we would hope they would be getting more accurate (that is, the error is going to zero), but more precisely, the truncation error implies that ratios of successive errors would be approaching 2. We demonstrate this in the examples below. In practice, a grid refinement study is critical to ensuring a method is implemented properly and should not be avoided. Unless you are getting the right ratio of successive errors it is not enough that the errors are going to zero.

In practice, how is h chosen? One might think that h close to zero would be best. From our previous chapter, we know there is also roundoff error present which, in computing the approximate derivative (3.3), is (roughly) proportional to $1/h$. The overall error therefore is of the form

$$E = ch + \frac{\delta}{h}$$

where c and δ are constants. This error function has the generic shape illustrated in Fig. 3.2. This places severe restriction on the accuracy that can be achieved with this formula, but it also provides insight.

For example, let's suppose that machine precision is $\bar{\epsilon} \approx 10^{-16}$. Then should $h = 10^{-16}$? Actually, no. To see this, note that our error is $\mathcal{O}(h + \frac{\bar{\epsilon}}{h})$ which is minimized when $h \approx \sqrt{\bar{\epsilon}} = 10^{-8}$.

Fig. 3.2 Error in numerical differentiation

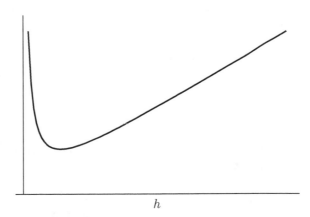

h

Example 3 Use Eq. (3.3) with $h = 2^{-1}, 2^{-2}, \ldots$ to estimate $f'(0.7)$ for $f(x) = \cos x$.

This is easily achieved with the Python commands:

```
>>> import numpy as np
>>> x0 = 0.7
>>> h = 2.**-np.arange(1,11)
>>> df = (np.cos(x0 + h) - np.cos(x0)) / h
```

The results produced are:

k	Approximation	Ratio of errors	Relative differences
1	**−0.80496886561563**	—	—
2	**−0.7**3263639128242	1.818068	0.0898574807
3	**−0.6**9028177328519	1.919472	0.0578112397
4	**−0.6**6769189557836	1.962328	0.0327255891
5	**−0.65**606252587900	1.981809	0.0174172695
6	**−0.650**16668229364	1.991065	0.0089867099
7	**−0.64**719878353456	1.995573	0.0045648275
8	**−0.645**70987940618	1.997797	0.0023005360
9	**−0.644**96419361632	1.998901	0.0011548310
10	**−0.6445**9104291166	1.999451	0.0005785603
	$\vdots$		
25	**−0.644217**69976616	1.892038	0.0000000173
26	**−0.6442176**9231558	2.467260	0.0000000116
27	**−0.644217**69976616	0.405308	0.0000000116
28	**−0.644217**69976616	1.000000	0.0000000000
29	**−0.64421**772956848	0.295966	0.0000000463

The true value is $f'(0.7) = -\sin 0.7 = -0.64421768723769102$ (to 17 digits) and the approximations to this for the values of $h = 2^{-k}$ are in the second column. The figures in **bold** type are the correct digits in each approximation. We see exactly the behavior predicted by Fig. 3.2, the error gets steadily smaller for the first several entries *and then begins to rise as h continues to decrease*. The next columns show the ratio of absolute errors between consecutive approximations

$$\left| \frac{\tilde{f}'_{k-1} - f'}{\tilde{f}'_k - f'} \right|,$$

where $\tilde{f}'_k$ is the kth approximation of f', and absolute relative difference between consecutive estimates

$$\left| \frac{\tilde{f}'_k - \tilde{f}'_{k-1}}{\tilde{f}'_{k-1}} \right|.$$

As the theory predicts (and described above) this ratio should be approximately
2 since we are halving h each time for this grid refinement study. This occurs until
the error begins to grow slightly towards the end.

We also see in Example 3 that we cannot get very high precision using this
formula. Similar results would be obtained using negative steps $-h$, called *backward
differences*. (See the exercises.)

Improved results can be obtained by using (3.3) with *both* h and $-h$ as follows:

$$f'(x_0) \approx \frac{f(x_0 + h) - f(x_0)}{h}$$

$$f'(x_0) \approx \frac{f(x_0 - h) - f(x_0)}{-h}$$

Averaging these approximations we get

$$
\begin{aligned}
f'(x_0) &\approx \frac{1}{2}\left[\frac{f(x_0 + h) - f(x_0)}{h} + \frac{f(x_0 - h) - f(x_0)}{-h}\right] \\
&= \frac{f(x_0 + h) - f(x_0 - h)}{2h}
\end{aligned}
\tag{3.5}
$$

Assuming that $f^{(3)}(x)$ is continuous in the interval $[x_0 - h, x_0 + h]$, we can
apply Taylor's theorem to each of $f(x_0 + h)$, $f(x_0 - h)$ (and the intermediate value
theorem) to obtain the truncation error

$$\frac{h^2}{6} f^{(3)}(\xi) = \mathcal{O}(h^2) \tag{3.6}$$

for some $\xi \in (x_0 - h, x_0 + h)$.

We will eventually see that the approximation (3.5) can also be obtained as the
result of differentiating the interpolation polynomial agreeing with f at the nodes
$x_0 - h$, x_0, and $x_0 + h$.

In Fig. 3.3, the differences between the two approaches is illustrated.

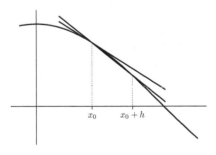

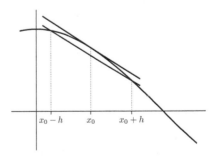

(a) Differing slopes of tangent and chord.   (b) Similar slopes of tangent and chord.

Fig. 3.3 Difference between the two numerical differentiation approaches (3.3) and (3.5)

Example 4 Estimate $f'(0.7)$ for $f(x) = \cos x$ using (3.5).

Using the same set of steplengths as in Example 3, we get

k	Approximation	Ratio of errors	Relative differences
1	**-0.61**770882336457	—	—
2	**-0.63**752802577759	3.962661	0.032085023984
3	**-0.64**254134718726	3.990635	0.007863687880
4	**-0.64**379835658979	3.997657	0.001956308972
5	**-0.644**11283921816	3.999414	0.000488480011
6	**-0.644**19147427288	3.999854	0.000122082731
7	**-0.64421**113393649	3.999963	0.000030518354
8	**-0.64421**604890865	3.999991	0.000007629443
9	**-0.64421**727765520	3.999998	0.000001907352
10	**-0.64421**758484207	4.000000	0.000000476837
	$\vdots$		
15	**-0.64421768**713873	4.042176	0.000000000467
16	**-0.64421768**721149	3.777420	0.000000000113
17	**-0.64421768**722968	3.271769	0.000000000028
18	**-0.64421768**723696	10.953827	0.000000000011
19	**-0.64421768**722241	0.047829	0.000000000023
20	**-0.64421768**725151	1.105777	0.000000000045
21	**-0.64421768**730972	0.191881	0.000000000090
22	**-0.64421768**719330	1.622751	0.000000000181
23	**-0.64421768**719330	1.000000	0.000000000000
24	**-0.64421768**765897	0.105363	0.000000000723
25	**-0.64421768**672764	0.825951	0.000000001446

Again, the correct figures are shown in **bold** type.

The accuracy improves more rapidly (by a factor 4 for each halving of h since the method is second order) and greater accuracy overall is achieved (11 decimal places for $k = 18$). Just as with the first-order approximation in Example 3, the accuracy is still limited and eventually the effect of the roundoff errors again takes over as we see the total error rising for small h.

Let's examine the impact of roundoff errors a little more closely. Consider the simple approximation (3.3). If h is small, then we can reasonably assume that $f(x_0)$ and $f(x_0 + h)$ have similar magnitudes, and, therefore, similar roundoff errors. Suppose these are bounded by δ, and denote the computed values of f by $\hat{f}$. Then the overall error is given by

$$
\left| f'(x_0) - \frac{\hat{f}(x_0 + h) - \hat{f}(x_0)}{h} \right|
$$

$$
\leq \left| f'(x_0) - \frac{f(x_0 + h) - f(x_0)}{h} \right| + \left| \frac{f(x_0 + h) - f(x_0)}{h} - \frac{\hat{f}(x_0 + h) - \hat{f}(x_0)}{h} \right|
$$

$$\leq \left| \frac{h}{2} f''(\xi) \right| + \left| \frac{f(x_0 + h) - \hat{f}(x_0 + h)}{h} \right| + \left| \frac{f(x_0) - \hat{f}(x_0)}{h} \right|$$

$$\leq \left| \frac{h}{2} f''(\xi) \right| + \frac{2\delta}{h}$$

Note that $\left| \frac{h}{2} f''(\xi) \right|$ is the truncation error in the approximation of the first derivative. It is the second term of this error bound which leads to the growth of the error as $h \to 0$ illustrated in Fig. 3.2. A similar effect is present for all numerical differentiation formulas.

Similar approaches to these can be used to approximate higher derivatives. For example, adding third order Taylor expansions for $f(x_0 \pm h)$, we obtain

$$f''(x_0) \approx \frac{f(x_0 + h) - 2f(x_0) + f(x_0 - h)}{h^2} \tag{3.7}$$

which has error given by $h^2 f^{(4)}(\theta)/12$. Similarly, the five point formula is

$$f''(x_0) \approx \frac{-f(x_0 + 2h) + 16f(x_0 + h) - 30f(x_0) + 16f(x_0 - h) - f(x_0 - 2h)}{12h^2} \tag{3.8}$$

which has truncation error of order h^4.

Keep in mind that the tendency for roundoff errors to dominate the computation of higher order derivatives is increased since the reciprocal power of h in the roundoff error estimate rises with the order of the derivative.

Example 5 Use Eqs. (3.7) and (3.8) to estimate $f''(0.7)$ for $f(x) = \cos x$.

Using (3.7):

$h = 0.1 \qquad f''(0.7) \approx 100 \left[\cos(0.8) - 2\cos(0.7) + \cos(0.6) \right] = -0.764\,205$

$h = 0.01 \qquad f''(0.7) \approx 10000 \left[\cos(0.71) - 2\cos(0.7) + \cos(0.69) \right] = -0.764\,836$

Using (3.8):

$$h = 0.1 \qquad f''(0.7) \approx -0.764\,841$$
$$h = 0.01 \qquad f''(0.7) \approx -0.764\,842$$

This last estimate is exact in all places shown.

Example 6 Putting it all together: Approximating heat flow on an insulated wire.

The derivatives in the model from Eq. (3.1) can be discretized using the above approximate derivatives and an explicit formula for the temperature on the wire at a given time can be derived. Consider using the forward difference approximation to the time derivative,

$$u_t = \frac{u(x, t + \Delta t) - u(x, t)}{\Delta t} + \mathcal{O}(\Delta t)$$

so that

$$u_t(x_i, t^k) \approx \frac{u_i^{k+1} - u_i^k}{\Delta t}.$$

For the second derivative term, using Eq. (3.7) gives,

$$u_{xx} = \frac{u(x - \Delta x, t) - 2u(x, t) + u(x + \Delta x, t)}{\Delta x^2} + \mathcal{O}(\Delta x^2)$$

and so

$$u_{xx} \approx \frac{u_{i-1}^k - 2u_i^k + u_{i+1}^k}{\Delta x^2}.$$

Putting this all together, $\rho c u_t - K u_{xx} = f$ can be approximated with the discrete model,

$$\rho c \left[\frac{u_i^{k+1} - u_i^k}{\Delta t} \right] - K \left[\frac{u_{i-1}^k - 2u_i^k + u_{i+1}^k}{\Delta x^2} \right] = f_i^k. \tag{3.9}$$

Note that, given the initial condition, $u(x, 0) = u_0(x)$ and the zero boundary conditions, we have everything we need to compute $u_i^1 \approx u(x_i, t^1)$ for $i = 0, 1, 2, \ldots n$. Then we would use those values to approximate temperatures at t^2. This formulation is called an *explicit finite difference approximation*. Explicit means you have a formula for calculating u_i^{k+1} in terms of u_i^k (i.e. only previous values). To see this, let $H = \frac{K \Delta t}{\rho c \Delta x^2}$ and solve for u_i^{k+1} to get

$$u_i^{k+1} = H u_{i-1}^k - (2H - 1)u_i^k + H u_{i+1}^k + \frac{\Delta t}{\rho c} f_i^k.$$

Some things to consider–how would you code this up? How would you perform a grid refinement study? See the exercises (and subsequent chapters) for more with this model.

Exercises

1. Use Eq. (3.3) with $h = 0.1, 0.01, \ldots, 10^{-6}$ to estimate $f'(x_0)$ for

    (a) $f(x) = \sqrt{x}$, $x_0 = 4$

(b) $f(x) = sin(x)$, $x_0 = \frac{\pi}{4}$

(c) $f(x) = \ln x$, $x_0 = 1$

2. Repeat Exercise 1 using negative steplengths $h = -10^{-k}$ for $k = 1, 2, \ldots, 6$.
3. Repeat Exercise 2 using $h = 2^{-k}$ for $k = 1, 2, \ldots 10$ and calculate the errors for each steplength. Calculate the ratio of consecutive errors and explain what you see.
4. Repeat Exercise 1 using (3.5). Compare the accuracy of your results with those of Exercises 1 and 2. Why are the results for (b) using (3.5) almost exact? Why are they *not exact*?
5. Estimate $f''(x_0)$ using both (3.7) and (3.8) for the same functions and steplengths as in the earlier exercises. Compare the accuracy of the two formulas.
6. Implement the computer program to approximate the temperatures along a wire as in Example 6 using numerical derivatives. Experiment with different values of the model parameters, including Δx and Δt. You will see that this approach can give accurate approximation but stability issues arise if $1 - 2\left(\frac{K\Delta t}{\rho c \Delta x^2}\right) > 0$. We will discuss ways to improve stability in Chap. 4 on Linear Equations.
7. The table below gives values of the height of a tomato plant in inches over days. Use these values to estimate the rate of growth of the tomato plant over time. Make some observations about what you see.

day	0	2	4	6	8	10	12	14	16	20
height (inches)	8.23	9.66	10.95	12.68	14.20	15.5	16.27	18.78	22.27	24.56

8. The table below shows the position of a drone being tracked by a sensor. Use this to tack its velocity and acceleration over time

t (s)	0	2	4	6	8	10	12
position (m)	0.7	1.9	3.7	5.3	6.3	7.4	8.0

9. More and more seafood is being farm-raised these days. A model used for the rate of change for a fish population, $P(t)$ in farming ponds is given by

$$P'(t) = b\left(1 - \frac{P(t)}{P_M}\right)P(t) - hP(t)$$

where b is the birth rate, P_M is the maximum number of fish the pond can support, and h is the rate the fish are harvested.

(a) Use forward differences to discretize this problem and write a script that takes in an initial population $P(0)$ as well as all model parameters and outputs (via a plot for example) the population over time.

(b) Demonstrate your model for a carrying capacity P_M of 20,000 fish, a birth
    rate of $b = 6\%$ and a harvesting rate of $h = 4\%$ if the pond starts with 5,000
    fish.
(c) Perform some numerical experiments to demonstrate how sensitive this
    resulting populations are to all of the model parameters in part (b).

10. The modification of an image can be modeled mathematically using numerical
    derivatives. The conceptual approach can be thought of in the context of diffu-
    sion; that is, high intensity pixels spread to low intensity pixels which can result
    in a picture that is less sharp than intended. Consider the 2-D diffusion model;

$$u_t - K u_{xx} - K u_{yy} = f(x, y, t) \tag{3.10}$$

Numerical derivatives can still be used to discretize this problem as in Example 6,
but you must consider approximations in both x and y. Consider uniform spacing
in both directions and let $h = \Delta x = \Delta y$ to simplify the notation. Now, let $u_{i,j}^k \approx$
$u(x_i, y_j, t^k)$ Here i and j indicate our location in space and k is our location
in time with $t^k = k\Delta t$, $x_i = i\Delta x$ and $y = j\Delta y$. For example our approximated
solution at the second location in x, third location in y at the 4th time step would
be $u_{2,3}^4$.
Following the derivations in Example 6 regarding a finite difference approximation to u_{xx} and a first derivative approximation in time, we can extend this to
include an approximation to u_{yy}. If we put it all together then our discrete model
can be written as the explicit finite difference scheme;

$$\left[\frac{u_{i,j}^{k+1} - u_{i,j}^k}{\Delta t} \right] = K \left[\frac{u_{i-1,j}^k + u_{i,j-1}^k - 4u_i^k + u_{i+1,j}^k + u_{i,j+1}^k}{h^2} \right] + f_i^k.$$

$$\tag{3.11}$$

Blurring of an image can be modeled by this explicit finite difference method
where the 2-D image is viewed as the initial "temperature" of the mass. The
reverse of heat diffusion is to use negative time steps and to start with the temperature distribution in the future. This suggests that the de-blurring of an image
may be modeled by starting with a blurred image as a future temperature and
apply the explicit finite difference model with negative times.
Provided here is an example based on Matlab code from Dr. Robert White at
N.C. State University to read in an image and take one iteration of blurring and
sharpening. It uses the picture Italy.jpg.

```python
import numpy as np
from skimage import io, color

# The new images will be saved in your current
# working directory
pic = io.imread('italy.jpg')
pic = color.rgb2gray(pic)
```

```
newpic = np.empty_like(pic)

coeff = 1

for j in range(1, pic.shape[1] - 1):
    for i in range(1, pic.shape[0] - 1):
        newpic[i, j] = (pic[i, j] + coeff *
            (pic[i - 1, j] - 2 * pic[i, j] +
            pic[i + 1, j]) + coeff *
            (pic[i, j - 1] - 2 * pic[i, j] +
            pic[i, j + 1]))

pic[1:-1,1:-1] = newpic[1:-1,1:-1]
# Scale image values to correct range for
# saving
pic /= np.max(np.abs(pic))

io.imsave('blurry.jpg', pic)
```

The results look like this with $K = 1$ (Figs. 3.4 and 3.5)

(a) Run this code using a picture of your choice by modifying the jpg file. Experiment with different time steps in the k-loop. Now consider different model coefficients. There is a stability requirement you may run in to and we hope you do! Summarize what you did and provide some illustrative pictures.

(b) Discussion: Explain in your own words, the implementation here in terms of the computation of newpic for blurring and sharpening. What is going on there? Specifically relate the code to the discretized model above. What is the meaning of coeff with regards to the heat flow (diffusion) model?

Fig. 3.4 Original image

Fig. 3.5 Resulting blurred
image

3.3 Numerical Integration

You already know more about numerical integration then you may think. Recall the
Riemann integral is defined as a limit of Riemann sums for arbitrary partitions over
the interval of integration. When you were first learning about integration, you likely
became familiar with the ideas by considering left and right endpoint approximations
for finding the area below a function and above the x axis. One could start by taking
uniform partitions with some number N of subdivisions. We can then denote the
steplength of these subdivisions by

$$h = \frac{b - a}{N}$$

and write

$$x_k = a + kh \qquad (k = 0, 1, \dots, N)$$

Then the Riemann left and right sums are given by

$$L = h \sum_{k=0}^{N-1} f(x_k) = h \sum_{k=0}^{N-1} f(a + kh) \tag{3.12}$$

$$R = h \sum_{k=1}^{N} f(x_k) = h \sum_{k=1}^{N} f(a + kh) \tag{3.13}$$

These Riemann sums are actually (very basic) numerical integration techniques.
They are equivalent to integrating step-functions which agree with f at the appropriate set of nodes. These step-functions are piecewise constant functions (i.e. degree
0 interpolating polynomials). The efficient methods we shall investigate result from
increasing the degree of the interpolating polynomial.

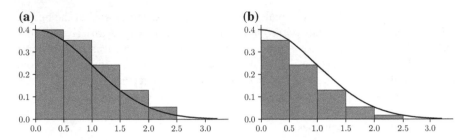

Fig. 3.6 Left (**a**) and right (**b**) sum approximations to an integral

Step function approximations to an integral are illustrated in Fig. 3.6. It is easy to see that the shaded areas will differ substantially from the true value of the integral. Clearly there is a need for better numerical integration methods than these.

Just as numerical derivatives could be approximated using only specific values of $f(x)$, numerical integration formulas can be constructed in the same way. Similarly, there are choices that must be made along the way in terms of achieving a desired accuracy. In this section, we introduce the ideas of numerical integration, in which we approximate a definite integral

$$I = \int_a^b f(x)\, dx$$

by a weighted sum of function values. That is, we use approximations of the form

$$\int_a^b f(x)\, dx \approx \sum_{k=0}^N c_k f(x_k) \tag{3.14}$$

where the *nodes* $x_0, x_1, \ldots, x_N \in [a, b]$ and the coefficients, or *weights*, c_k are chosen appropriately. This is often referred to as an interpolatory quadrature formula.

Note that Eq. (3.14) is actually a particular form of a Riemann sum where the nodes are the representative points from the various subintervals in the partition. The weights would be the widths of the subdivisions of the interval. Clearly as a user of numerical integration, algorithms should be accurate in that they give approximations of the true integrals but also simplicity is desirable. All the methods we shall consider are based on approximating $f(x)$ with a polynomial $p(x)$ on each subinterval and integrating the polynomial exactly;

$$\int_a^b f(x)\, dx \approx \int_a^b p(x)\, dx.$$

Most importantly, the polynomials used have the same values as f at the nodes. This is called an *interpolating polynomial*. These ideas can be used to derive specific nodes and weights that ensure a particular accuracy, resulting in a technique of form in Eq. (3.14), which is just a weighted sum of function values. It is important to note here, the numerical integration schemes will be derived in terms of function values of $f(x)$ at the nodes, but in practice, data values could be used instead.

The following is a definition that we will use repeatedly in this section to derive numerical integration schemes.

Definition 1 The numerical integration formula (3.14) has *degree of precision m* if it is exact for all polynomials of degree at most m, but is not exact for x^{m+1}.

The key idea is that any interpolatory quadrature rule using the nodes $x_0, x_1, \ldots, x_N$ must have degree of precision *at least* N since it necessarily integrates polynomials of degree up to N exactly. (We shall see later that the degree of precision can be greater than N.) We illustrate the process with an example.

Example 7 Find the quadrature rule using the nodes $x_0 = 0$, $x_1 = 1$, and $x_2 = 2$ for approximating $\int_0^3 f(x)\,dx$. Use the resulting rule to estimate $\int_0^3 x^3 dx$.

Using three nodes, we expect a degree of precision of at least 2. Thus, we seek weights c_0, c_1, c_2 such that

$$I = \int_0^3 f(x)\,dx = c_0 f(0) + c_1 f(1) + c_2 f(2)$$

for $f(x) = 1, x, x^2$. Now, with

$f(x) = 1$	$I = 3$	$c_0 f(0) + c_1 f(1) + c_2 f(2) = c_0 + c_1 + c_2$
$f(x) = x$	$I = 9/2$	$c_0 f(0) + c_1 f(1) + c_2 f(2) = c_1 + 2c_2$
$f(x) = x^2$	$I = 9$	$c_0 f(0) + c_1 f(1) + c_2 f(2) = c_1 + 4c_2$

Solving the equations

$$c_0 + c_1 + c_2 = 3, \quad c_1 + 2c_2 = 9/2, \quad c_1 + 4c_2 = 9$$

we obtain

$$c_0 = \frac{3}{4}, \quad c_1 = 0, \quad c_2 = \frac{9}{4}$$

The required quadrature rule is therefore

$$\int_0^3 f(x)\,dx \approx \frac{3f(0) + 0f(1) + 9f(2)}{4}$$

Now, with $f(x) = x^3$, we get

$$\int_0^3 x^3 dx \approx \frac{3(0)^3 + 0(1)^3 + 9(2)^3}{4} = \frac{72}{4}$$

whereas the true integral is $3^4/4 = 81/4$. This shows that the degree of precision of this formula is 2 since it fails to be exact for x^3. Actually, for $f(x) = x^3$, the error is $9/4$.

Note though, that the scheme should integrate any second degree polynomial exactly. For example, consider $f(x) = 3x^2 + 2x + 1$ so that

$$\int_0^3 f(x)dx = 39$$

while the quadrature rule gives

$$\frac{3f(0) + 0f(1) + 9f(2)}{4} = \frac{3(1) + 0(6) + 9(17)}{4} = 39.$$

In Example 7, we set out to find an integration scheme with degree of precision at least two, and the procedure was to find weights so that the rule is exact for the functions $1, x, x^2$. The fact that this is sufficient to establish exactness for all quadratics follows readily from the linearity of both the integral and the numerical integration formula. Thus, for any quadratic polynomial $p(x) = \alpha x^2 + \beta x + \gamma$, we have

$$\int_a^b p(x)\,dx = \alpha \int_a^b x^2 dx + \beta \int_a^b x dx + \gamma \int_a^b 1 dx$$

and a similar result holds for the quadrature rule.

It is clear that, following the same procedure used in Example 7, we can find the weights for the interpolatory quadrature rule using any set of nodes by solving a system of linear equations. We present three standard integration formulas which, although simplistic, are used in practice and often embedded in powerful, large-scale simulation tools (Figs. 3.7, 3.8 and 3.9).

The Midpoint Rule

$$\int_a^b f(x)\,dx \approx (b - a)\, f((a + b)/2) \tag{3.15}$$

The Trapezoid Rule

$$\int_a^b f(x)\,dx \approx \frac{b - a}{2}\,[f(a) + f(b)] \tag{3.16}$$

Fig. 3.7 Midpoint quadrature rule

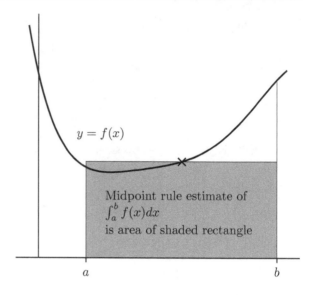

Midpoint rule estimate of $\int_a^b f(x)dx$ is area of shaded rectangle

Fig. 3.8 Trapezoid quadrature rule

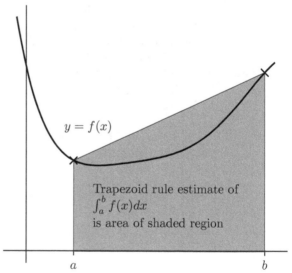

Trapezoid rule estimate of $\int_a^b f(x)dx$ is area of shaded region

Simpson's Rule

$$\int_a^b f(x)\,dx \approx \frac{b-a}{6}\left[f(a) + 4f\left((a+b)/2\right) + f(b)\right] \qquad (3.17)$$

Fig. 3.9 Simpson's
quadrature rule

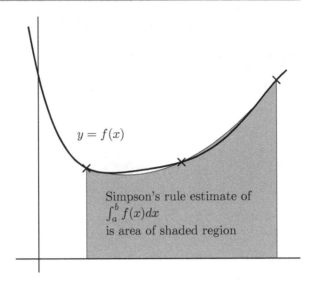

$y = f(x)$

Simpson's rule estimate of
$\int_a^b f(x)\,dx$
is area of shaded region

Example 8 Estimate the integral for the standard normal distribution function

$$N_{0,1}\,(1/2) = \frac{1}{2} + \frac{1}{\sqrt{2\pi}} \int_0^{1/2} \exp\left(-x^2/2\right) dx$$

(a) Using the midpoint rule:

$$N_{0,1}\,(1/2) \approx \frac{1}{2} + \frac{1}{\sqrt{2\pi}}\frac{1}{2}\exp\left(-\left(1/4\right)^2/2\right) \approx 0.693334$$

(b) Using the trapezoid rule

$$N_{0,1}\,(1/2) \approx \frac{1}{2} + \frac{1}{\sqrt{2\pi}}\frac{1/2}{2}\left[\exp\left(0\right) + \exp\left(-\left(1/2\right)^2/2\right)\right] \approx 0.687752$$

(c) Using Simpson's rule

$$N_{0,1}\,(1/2) \approx \frac{1}{2} + \frac{1}{\sqrt{2\pi}}\frac{1/4}{3}\left[\exp\left(0\right) + 4\exp\left(-\left(1/4\right)^2/2\right) + \exp\left(-\left(1/2\right)^2/2\right)\right]$$
$$\approx 0.691473$$

The true value is 0.6914625 to 7 decimals. The errors in these approximations are therefore approximately 0.0019, 0.0037 and 0.00001 respectively. It is obvious that Simpson's rule has given a significantly better result. We also see that the error in the midpoint rule is close to one-half that of the trapezoid rule, which will be explained soon!

These schemes can be obtained by integrating the interpolation polynomials which agree with f at the given nodes, or by solving the appropriate linear systems. We provide the derivations with the special case of Simpson's rule for the interval $[-h, h]$.

We require an integration rule using the nodes $-h, 0, h$ which is exact for all quadratics. Therefore, we must find coefficients c_0, c_1, c_2 such that

$$I = \int_{-h}^{h} f(x)\, dx = c_0 f(-h) + c_1 f(0) + c_2 f(h)$$

for $f(x) = 1, x, x^2$. These requirements yield the equations

$$c_0 + c_1 + c_2 = 2h$$
$$-c_0 h + c_2 h = 0$$
$$c_0 (-h)^2 + c_2 h^2 = \frac{2h^3}{3}$$

The second of these immediately gives us $c_0 = c_2$, and then the third yields $c_0 = h/3$. Substituting these into the first equation, we get $c_1 = 4h/3$ so that Simpson's rule for this interval is

$$\int_{-h}^{h} f(x)\, dx \approx \frac{h}{3} [f(-h) + 4f(0) + f(h)]$$

For a general interval $[a, b]$, we can use $h = (b - a)/2$ and $m = (a + b)/2$ with the change of variables

$$x = \frac{a + b}{2} + s$$

to get

$$\int_{a}^{b} f(x)\, dx = \int_{-h}^{h} f(m + s)\, ds \approx \frac{h}{3} [f(a) + 4f(m) + f(b)]$$

which corresponds to Eq. (3.17).

The derivations of the midpoint and trapezoid rules are left as exercises.

Simpson's rule is a case in which the quadrature rule has a higher degree of precision than that of the interpolation polynomial; it is exact not just for quadratics but for cubics, too. To see this consider applying Simpson's rule to $\int_{-h}^{h} x^3 dx$. The integral evaluates to 0, since it is an integral of an odd function over a symmetric interval. Simpson's rule gives the approximation

$$\frac{h}{3} \left[(-h)^3 + 4(0)^3 + h^3 \right] = -\frac{h^4}{3} + \frac{h^4}{3} = 0$$

which is exact, as claimed. You can show that Simpson's rule is not exact for $f(x) = x^4$. Similarly, the midpoint rule gives higher precision: it has degree of precision 1.

In practice, what matters is how well a numerical integration scheme performs on a general integral, not on a polynomial (since any student who has had Calculus could integrate a polynomial exactly without using numerical integration!) Luckily, error formulas for interpolatory quadrature rules have been derived. The expressions can be obtained by integrating the so-called Lagrange interpolation remainder term, but those derivations are beyond the scope of this text. Later, we shall, however, give some reasons to believe the results quoted below for the errors in the midpoint, trapezoid and Simpson's rules:

$$\int_a^b f(x)\,dx - (b-a)f((a+b)/2) = \frac{(b-a)h^2}{24}f''(\xi_M) \qquad (3.18)$$

where $h = b - a$,

$$\int_a^b f(x)\,dx - \frac{b-a}{2}[f(a) + f(b)] = -\frac{(b-a)h^2}{12}f''(\xi_T) \qquad (3.19)$$

where again $h = b - a$, and

$$\int_a^b f(x)\,dx - \frac{h}{3}[f(a) + 4f(a+h) + f(b)] = -\frac{(b-a)h^4}{180}f^{(4)}(\xi_S) \qquad (3.20)$$

where $h = (b-a)/2$. Here the points ξ_M, ξ_T, ξ_S are "mean value points" in the interval (a, b).

The reason for the, apparently unnecessary, use of $h = b - a$ in (3.18) and (3.19) will become apparent in the next section when we look at the "composite" forms of these integration rules. We will also demonstrate the usefulness of these error estimates.

It sometimes puzzles students that the midpoint rule is more precise than the trapezoid rules since it seems like the trapezoid rule is using a more approximate representation of the area below the curve. One might wonder, why even bother having the trapezoid rule at all if the midpoint rule seems simpler and is more accurate. This is a completely reasonable question. One answer is that in practice, scientists and engineers are often working solely with data, not an analytic function. In that case, they only have access to endpoints of intervals, not midpoints.

We have seen that interpolatory quadrature rules can have degree of precision greater than we should expect from the number of nodes. We next consider choosing the nodes themselves, as well as the appropriate weights. This is a new perspective in that we now seek the quadrature rule with maximum degree of precision for a given number of nodes. Such an approach to numerical integration is known as *Gaussian quadrature*. The general theory of Gaussian integration is beyond the scope of the present text. However we demonstrate the idea with the following examples which step through the derivation process and also compare to the previous schemes.

Example 9 Find the quadrature formulas with maximum degree of precision, the Gaussian rules, on $[-1, 1]$ using

(a) two points, and
(b) three points

For simplicity, denote $\int_{-1}^{1} f(x)\,dx$ by $I(f)$.

(a) We now seek the nodes x_0, x_1 and their weights c_0, c_1 to make the integration rule exact for $f(x) = 1, x, x^2, \ldots$ as high of a degree polynomial as possible. Since there are four unknowns, it makes sense to try to satisfy four equations. Note that

$$I\left(x^k\right) = \int_{-1}^{1} x^k dx = \begin{cases} \frac{2}{k+1} \text{ if } k \text{ is even} \\ 0 \text{ if } k \text{ is odd} \end{cases}$$

Thus we need:

(i)   For $f(x) = 1$, $I(f) = 2 = c_0 + c_1$
(ii) For $f(x) = x$, $I(f) = 0 = c_0 x_0 + c_1 x_1$
(iii) For $f(x) = x^2$, $I(f) = \frac{2}{3} = c_0 x_0^2 + c_1 x_1^2$
(iv) For $f(x) = x^3$, $I(f) = 0 = c_0 x_0^3 + c_1 x_1^3$

We can assume that neither c_0 nor c_1 is zero since we would then have just a one-point formula. From (ii), we see that $c_0 x_0 = -c_1 x_1$, and substituting this in (iv), we get

$$0 = c_0 x_0^3 + c_1 x_1^3 = c_0 x_0 \left(x_0^2 - x_1^2\right)$$

We have already concluded that $c_0 \neq 0$. If $x_0 = 0$, then (ii) would also give us $x_1 = 0$ and again the formula would reduce to the midpoint rule. It follows that $x_0^2 = x_1^2$, and therefore that $x_0 = -x_1$. Now (ii) implies that $c_0 = c_1$, and from (i), we deduce that $c_0 = c_1 = 1$. Substituting all of this into (iii), we get $2x_0^2 = 2/3$ so that we can take

$$x_0 = -\frac{1}{\sqrt{3}}, \quad x_1 = \frac{1}{\sqrt{3}}$$

So the resulting integration rule is

$$\int_{-1}^{1} f(x)\,dx \approx f\left(-1/\sqrt{3}\right) + f\left(1/\sqrt{3}\right) \tag{3.21}$$

One can deduce that $x_0 = -x_1$ and $c_0 = c_1$.

(b) Next, we want to find three nodes x_0, x_1, x_2 and their associated weights c_0, c_1, c_2. If the nodes are ordered in their natural order, then symmetry implies that

$$x_0 = -x_2, \quad x_1 = 0$$
$$c_0 = c_2$$

Using these facts in the equations for $f(x) = 1, x^2, x^4$, we get

$$2c_0 + c_1 = 2$$
$$2c_0 x_0^2 = \frac{2}{3}$$
$$2c_0 x_0^4 = \frac{2}{5}$$

(The equations for $f(x) = x, x^3, x^5$ are automatically satisfied with $x_0 = -x_2, x_1 = 0$ and $c_0 = c_2$.) The last pair of these gives $x_0^2 = 3/5$ and then $c_0 = 5/9$, leading to $c_1 = 8/9$. Thus, the Gaussian quadrature using three nodes in $[-1, 1]$ is given by

$$\int_{-1}^{1} f(x)\, dx \approx \frac{1}{9}\left[5f\left(-\sqrt{3/5}\right) + 8f(0) + 5f\left(\sqrt{3/5}\right)\right] \qquad (3.22)$$

and this has degree of precision 5.

Example 10 Use the Gaussian rules with two and three nodes to estimate

$$\int_{-1}^{1} \frac{dx}{x+2}$$

Compare the results with those obtained using the midpoint, trapezoid and Simpson's rules.

For comparison, let's calculate the midpoint, trapezoid and Simpson's rule estimates:

$$M = 2\left(\frac{1}{0+2}\right) = \frac{1}{2}$$
$$T = 1\left[\frac{1}{(-1)+2} + \frac{1}{1+2}\right] = 1\frac{1}{3} = 1.33333$$
$$S = \frac{1}{3}\left[1 + 4\left(\frac{1}{2}\right) + \frac{1}{3}\right] = \frac{10}{9} \approx 1.11111$$

Note the true value is $ln(3) \approx 1.09861$. The two Gaussian rules obtained in Example 9 yield

$$(3.21): \quad G_2 = \frac{1}{\left(-1/\sqrt{3}\right)+2} + \frac{1}{\left(1/\sqrt{3}\right)+2} = 1.09091$$

$$(3.22): \quad G_3 = \frac{1}{9}\left[\frac{5}{(-\sqrt{3/5})+2} + 8\left(\frac{1}{2}\right) + \frac{5}{\sqrt{3/5}+2}\right] \approx 1.09804$$

The Gaussian rules have significantly greater accuracy than the simpler rules using similar numbers of nodes.

Exercises

1. Derive the formula for the basic trapezoid rule by finding coefficients so that

$$\int_a^b f(x)\,dx = \alpha f(a) + \beta f(b)$$

   for $f(x) = 1, x$.
2. Find the quadrature rule for the interval $[-2, 2]$ using the nodes $-1, 0, +1$. What is its degree of precision?
3. Repeat Exercise 2 for the nodes $-2, -1, 0, +1, +2$. Use the resulting formula to estimate the area of a semicircle of radius 2.
4. Use the midpoint, trapezoid and Simpson's rules to estimate $\int_0^1 \frac{1}{1+x^2}\,dx$. Compare the results to the true value $\pi/4$.
5. Use the midpoint, trapezoid and Simpson's rules to estimate $\int_1^2 \frac{1}{x}\,dx$. Compare the results to the true value $\ln 2$.
6. Find the Gaussian quadrature rule using three nodes in $[0, 1]$. Use this rule to estimate $\int_0^1 \frac{1}{1+x^2}\,dx$ and $\int_0^1 \frac{1}{x+1}\,dx$. Compare the results with those obtained in Exercises 4 and 5.

3.4 Composite Formulas

The accuracy obtained from the trapezoid rule (or any other quadrature formula) can be improved by using the fact that

$$\int_a^b f(x)\,dx = \int_a^c f(x)\,dx + \int_c^b f(x)\,dx$$

for some intermediate point $c \in (a, b)$. For example, using $c = 1/4$ for the trapezoid rule in Example 8 we get

$$N_{0,1}(1/2) \approx \frac{1}{2} + \frac{1}{\sqrt{2\pi}}\frac{1/4}{2}\left[\exp(0) + \exp\left(-(1/4)^2/2\right)\right]$$

$$+ \frac{1}{\sqrt{2\pi}}\frac{1/4}{2}\left[\exp\left(-(1/4)^2/2\right) + \exp\left(-(1/2)^2/2\right)\right]$$

$$= 0.690543$$

reducing the error to about 0.0009 (See Fig. 3.10). Notice that by cutting the steplength h, in half we have reduced the error to approximately one-fourth of its previous value. This is directly related to our error expression from Eq. (3.19) and our

Fig. 3.10 Trapezoid applied to interval in two halves

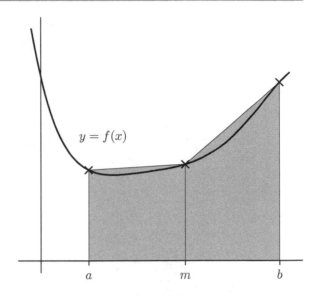

$$y = f(x)$$

previous notion of the grid refinement study presented in the context of numerical derivatives. We'll discuss this more soon.

We can rewrite the formula for the trapezoid rule using two subdivisions of the original interval to obtain

$$\int_a^b f(x)\,dx \approx \frac{h}{2}\left[f(a) + f(a+h)\right] + \frac{h}{2}\left[f(a+h) + f(b)\right] \quad (3.23)$$

$$= \frac{h}{2}\left[f(a) + 2f(a+h) + f(b)\right]$$

where now $h = (b-a)/2$. Geometrically, it is easy to see that this formula is just the average of the left and right sums for this partition of the original interval. Algebraically, this is also easy to see: Eq. (3.23) is exactly this average since the corresponding left and right sums are just $h\left[f(a) + f(a+h)\right]$ and $h\left[f(a+h) + f(b)\right]$.

This provides an easy way to obtain the general formula for the *composite trapezoid rule* T_N using N subdivisions of the original interval. With $h = (b-a)/N$, the left and right sums are given by (3.12) and (3.13), and averaging these we get

$$\int_a^b f(x)\,dx \approx T_N = \frac{1}{2}(L_N + R_N)$$

$$= \frac{1}{2}\left[h \sum_{k=0}^{N-1} f(a+kh) + h \sum_{k=1}^{N} f(a+kh)\right]$$

$$= \frac{h}{2}\left[f(a) + 2\sum_{k=1}^{N-1} f(a+kh) + f(b)\right] \quad (3.24)$$

Program Python implementation of composite trapezoid rule

```python
import numpy as np

def trapsum(fcn, a, b, N):
    """
    Function for approximating the integral of
    the function 'fcn' over the interval [a, b]
    in N segments using the trapezoid rule.

    """

    h = (b - a) / N
    s = (fcn(a) + fcn(b)) / 2
    for k in range(1, N):
        s += fcn(a + k * h)
    return s * h
```

Here s accumulates the sum of the function values which are finally multiplied by the steplength h to complete the computation. In the version above the summation is done using a loop. The explicit loop can be avoided by using Python's sum function as follows.

```python
def trapN(fcn, a, b, N):
    """
    Computes the trapezoid rule for the integral
    of the function 'fcn' over [a, b] using N
    subdivisions.

    """

    h = (b - a) / N
    x = a + np.arange(1, N) * h
    return h * ((fcn(a) + fcn(b)) / 2 + sum(fcn(x)
        ))
```

For the latter version it is essential that the function fcn is written for vector inputs.

Let's take a look back at our error formulas. It makes sense that the error formula (3.19) remains valid for the composite trapezoid rule – except with $h = (b - a) / N$. The proof of this result is left as an exercise while we provide the proof for Simpson's rule later.

Next, consider this error formula empirically. The error should depend in some way on the function to be integrated. Since the trapezoid rule integrates polynomials of degree 1 exactly, it is reasonable to suppose that this dependence is on the second derivative. (After all $f''(x) \equiv 0$ for any polynomial of degree 1.) Furthermore, if a function f varies slowly, we expect $\int_a^b f(x)\,dx$ to be approximately proportional to the length of the interval $[a, b]$. Using a constant steplength, the trapezoid rule estimate of this integral is expected to be proportional to $b - a$. In that case the error should be, too. That the error depends on the steplength seems natural and so we may reasonably suppose that the error

$$E(T_N) = \int_a^b f(x)\,dx - T_N = c\,(b - a)\,h^k f''(\xi)$$

for some "mean value point" ξ, some power k, and some constant c.

We set $f(x) = x^2$ to achieve a constant f''. By fixing $[a, b]$, we can determine suitable k and c values. $[a, b] = [0, 3]$ is a convenient choice so that $\int_a^b f(x)\,dx = 9$. The following Python loop generates a table error values $E(T_N)$ for this integral with $N = 1, 2, 4, 8$.

```
for k in range(4):
    N = 2**k
    I = trapN(lambda x: x**2, 0, 3, N)
    print(N, 9-I)
```

The results, and the ratios of successive errors, are

N	$E(T_N)$	$E(T_N)/E(T_{2N})$
1	-4.5	4.0
2	-1.125	4.0
4	-0.28125	4.0
8	-0.0703125	

For each time the number of subdivisions is doubled, i.e. h is halved, the error is reduced by a factor of 4. This provides strong evidence that $k = 2$. Finally, for $N = 1$, it is seen that $E(T_N) = -4.5$ and $b - a = 3, h = 3$, and $f''(\xi) = 2$. This means that the constant c must satisfy

$$c\,(3)\,(3^2)\,(2) = -4.5$$

From this we conclude that $c = -1/12$ which was stated in (3.19).

While this is by no means a proof of the error formula, this example demonstrates that the error formula does indeed hold and in addition, shows how to use those error expression to validate your code. If the ratio of successive errors was not 4 for a simple problem as above, then that is a red flag that there is a bug somewhere in your scripts. To this end, it is critical to know how to use such error expressions even if there is only a sense of understanding the theory behind them (which can be

Fig. 3.11 Nodes used by the composite trapezoid and midpoint rules

complicated). Similar arguments can be given in support of the other error formulas
(3.18) and (3.20) for the midpoint and Simpson's rules.

The composite midpoint rule uses the midpoints of each of the subdivisions. Let
$x_k = a + kh$ in the trapezoid rule and then in the midpoint rule we can use the same
N subdivisions of $[a, b]$ with the nodes $y_k = (x_{k-1} + x_k)/2 = a + \left(k - \frac{1}{2}\right) h$ for
$k = 1, 2, \ldots, N$. This gives

$$\int_a^b f(x)\, dx \approx M_N = h \sum_{k=1}^{N} f(y_k) = h \sum_{k=1}^{N} f\left(a + \left(k - \tfrac{1}{2}\right) h\right) \tag{3.25}$$

Figure 3.11 shows the distribution of nodes for the midpoint and trapezoid rules. The
composite version of Simpson's rule using $h = (b - a)/2N$ uses all the points for
the composite midpoint rule (3.25) as and composite trapezoid rule (3.24) together.
Rewriting these two rules with this value of h, we get

$$M_N = 2h \sum_{k=1}^{N} f(y_k) = h \sum_{k=1}^{N} f(a + (2k - 1) h)$$

$$T_N = h \left[f(a) + 2 \sum_{k=1}^{N-1} f(x_k) + f(b) \right] = h \left[f(a) + 2 \sum_{k=1}^{N-1} f(a + 2kh) + f(b) \right]$$

Now applying Simpson's rule to each of the intervals $\left[x_{k-1}, x_k\right]$ we get the composite
Simpson's rule formula

$$S_{2N} = \frac{h}{3} \left[\begin{array}{c} f(a) + 4f(a + h) + 2f(a + 2h) + \cdots \\ +2f(a + 2(N-1)h) + 4f(a + (2N-1)h) + f(b) \end{array} \right]$$

$$= \frac{h}{3} \left[f(a) + 4 \sum_{k=1}^{N} f(y_k) + 2 \sum_{k=1}^{N-1} f(x_k) + f(b) \right] \tag{3.26}$$

$$= \frac{T_N + 2M_N}{3} \tag{3.27}$$

The formula in (3.27) can be used to create an efficient implementation of Simpson's rule. it is important to note that Simpson's rule is only defined for an even
number of subdivisions, since each subinterval on which Simpson's rule is applied
must be subdivided into two pieces.

The following code implements this version using explicit loops. As with the trapezoid rule, the sum function can be used to shorten the code.

Program Python code for composite Simpson's rule using a fixed number of subdivisions

```python
import numpy as np

def simpsum(fcn, a, b, N):
    """
    Function for approximating the integral of
    the function 'fcn' over the interval [a, b]
    in N segments using the Simpson rule.

    """

    h = (b - a) / N
    s = (fcn(a) + fcn(b))
    for k in range(1, N, 2):
        s += 4 * fcn(a + k * h)
    for k in range(2, N - 1, 2):
        s += 2 * fcn(a + k * h)

    return s * h / 3
```

It would be easy to reinitialize the sum so that only a single loop is used, but the structure is kept simple here to reflect (3.26) as closely as possible.

Example 11 Use the midpoint, trapezoid and Simpson's rules to estimate

$$\int_{-1}^{1} \frac{dx}{x+2}$$

using $N = 1, 2, 4, 8, 16$ subdivisions. The true value is $I = ln(3) \approx 1.09861$. Compare the errors in the approximations.

Using the trapsum and Simpsum functions, and a similar program for the midpoint rule, we obtain the following table of results:

N	T_N	M_N	S_N	$E(T_N)$	$E(M_N)$	$E(S_N)$
1	1.3333	1.0000		0.2347	-0.0986	
2	1.1667	1.0667	1.11111	0.0681	-0.0319	0.012499
4	1.1167	1.0898	1.10000	0.0181	-0.0089	0.001388
8	1.1032	1.0963	1.09873	0.0046	-0.0023	0.000113
16	1.0998	1.0980	1.09862	0.0012	-0.0006	0.000008

It may not be surprising that Simpson's rule estimates are *much* more accurate for the same numbers of subdivisions – and that their errors decrease faster.

Next, we provide the theorem regarding errors in composite Simpson's rule estimates. This result will also be helpful in our practical algorithm in the next section.
This theorem establishes that the error formula (3.20) for the basic Simpson's rule
remains valid for the composite rule.

Theorem 2 *Suppose the integral* $I = \int_a^b f(x)\,dx$ *is estimated by Simpson's rule*
S_N *using* N *subdivisions of* $[a, b]$ *and suppose that* $f^{(4)}$ *is continuous. Then the*
error in this approximation is given by

$$I - S_N = -\frac{(b-a)\,h^4}{180} f^{(4)}(\xi) \tag{3.28}$$

where $h = (b-a)/N$ *for some* $\xi \in (a, b)$.

Proof First we note that N is necessarily even, and we denote $N/2$ by M. The
composite Simpson rule is equivalent to applying the basic Simpson's rule to each of
the M intervals $[x_k, x_{k+1}]$ where $x_k = a + 2kh$ $(k = 0, 1, \ldots, M - 1)$. By (3.20)

$$\int_{x_k}^{x_{k+1}} f(x)\,dx - \frac{h}{3}\left[f(x_k) + 4f(x_k + h) + f(x_{k+1})\right] = -\frac{(x_{k+1} - x_k)\,h^4}{180} f^{(4)}(\xi_k)$$

for some $\xi_k \in (x_k, x_{k+1})$.

Since $x_{k+1} - x_k = 2h$, it follows that the total error is given by

$$E(S_N) = -\frac{2h^5}{180} \sum_{k=0}^{M-1} f^{(4)}(\xi_k)$$

Because this fourth derivative is continuous, and therefore bounded, it follows from
the intermediate value theorem that there exists a point $\xi \in (a, b)$ such that

$$M f^{(4)}(\xi) = \sum_{k=0}^{M-1} f^{(4)}(\xi_k)$$

and so we may write

$$E(S_N) = -\frac{2Mh^5}{180} f^{(4)}(\xi)$$

Finally, we note that $2Mh = Nh = (b - a)$ so that

$$E(S_N) = -\frac{(b-a)\,h^4}{180} f^{(4)}(\xi)$$

as required.   ∎

Example 12 Obtain an error bound for the Simpson's rule estimates found in Example 11

It is straightforward to show that for $f(x) = 1/(x+2)$ on $[-1, 1]$, the fourth derivative is bounded by 24. That is, for $x \in [0, 1]$, we have

$$\left| f^{(4)}(x) \right| \le 24$$

In each case, $b - a = 2$ and $h^4 = 1/N^4$. Therefore, with $N = 2, 4, 8, 16$, we obtain error bounds

$$\frac{24(16)}{180N^4} = \left(\frac{384}{180\,(2^4)}, \frac{384}{180\,(4^4)}, \frac{384}{180\,(8^4)}, \frac{384}{180\,(16^4)} \right)$$
$$\approx \left(1 \times 10^{-1}, 8 \times 10^{-3}, 5 \times 10^{-4}, 3 \times 10^{-5} \right)$$

Note that these error bounds are all significantly larger than the actual errors. In practice, you don't know the exact error, but the Theorem 2 can still be used as the basis of practical numerical integration routines since it provides bounds.

Up until now, our discussion and analysis assumes we have access to a function $f(x)$. However, one of the powerful aspects of numerical calculus is that the approximations rely only on function *values*. To this end an analytic function is not needed and these methods can be used with data alone (which is often the case in the real-world). The following example shows this and we leave it as an exercise to modify the numerical integration subroutines in this chapter to handle data instead of functions as input.

Example 13 Suppose a population of beetles changed at the rate of r(t) beetles/week as shown below in Fig. 3.12. Use the trapezoid rule with six subintervals to estimate the area below the curve and interpret the results (i.e. what does your answer mean?)

With $N = 6$, $h = \frac{b-a}{n} = 4$ and the approximation is

$$\int_0^{24} r(t)dt \approx \frac{4}{2} [r(0) + 2r(4) + 2r(8) + 2r(12) + 2r(16) + 2r(20) + r(24)].$$

However, we do not have an analytic expression for $r(t)$ and therefore need to estimate those values based on the curve. There is inevitably human error that cannot be avoided but with a careful eye, suppose the follow values are taken in our sum:

$$2\,[0 + 2(500) + 2(3000) + 2(11000) + 2(3000) + 2(500) + 0] = 72000$$

Fig. 3.12 Plot of rate of
change of beetle population
over time

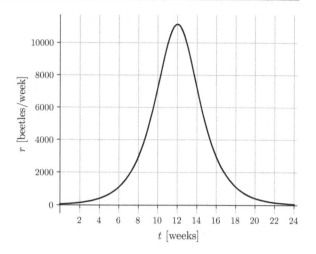

Since we are integrating the population rate, the solution here is the total beetle
population.

Exercises

1. Write a program for the midpoint rule using N subdivisions for $\int_a^b f(x)\,dx$.
   Use it to find estimates of $\ln 5 = \int_1^5 (1/x)\,dx$ using $N = 1, 2, 4, 8, 16, 32$.
2. Use your answers for Exercise 1 to verify that the error in the midpoint rule is
   proportional to h^2.
3. Derive the error formula for the composite trapezoid rule. Use this formula to
   obtain bounds on the trapezoid rule errors in Example 11.
4. Repeat Exercises 1 and 2 for the trapezoid rule.
5. Repeat the computation of Exercise 1 for Simpson's rule with $N=2, 4, 8, 16, 32$.
   Use the results to check that the error is proportional to h^4.
6. Use the midpoint, trapezoid and Simpson's rules to estimate

$$N_{0,1}(2) = \frac{1}{2} + \frac{1}{\sqrt{2\pi}} \int_0^2 \exp\left(-x^2/2\right) dx$$

   using $N = 10$ subdivisions.
7. A model for the shape of a lizard's egg can be found by rotating the graph
   of $y = (ax^3 + bx^2 + cx + d)\sqrt{1 - x^2}$ around the x-axis. Consider using $a =
   -0.05, b = 0.04, c = 0.1, d = 0.48$. Set up an integral to find the volume of
   the resulting solid of revolution (you may need to consult a Calculus text to
   remember the techniques to do that!). You may want to plot the model equation
   on the interval $[-1,1]$. Use numerical integration to approximate on the resulting
   integral to approximate the volume.
8. Duplicate the results in Example 13 by modifying your trapezoid code to work
   with data instead of function evaluations.

9. Modify your Simpson's rule code to work with data instead of function evaluations and apply it to the problem in Example 13 using $N = 6$ subintervals. Increase the number of subintervals. How does this impact your solution?
10. A sensor was used to track the speed of a bicyclist during the first 5 s of a race. Use numerical integration to determine the distance she covered.

t(s)	0	0.5	1.0	1.5	2.0	2.5	3.0	3.5	4.0	4.5	5.0
v (m/s)	0	7.51	8.10	8.93	9.32	9.76	10.22	10.56	11.01	11.22	11.22

11. Just to get you outside (and to gain a better understanding of numerical integration), your mission is to find a leaf and using composite numerical integration, approximate the area of the leaf using two numerical integration techniques of your choice. Use at least 10 subintervals.

    One approach to this would be to trace the leaf aligned on a coordinate system and measure different points on the leaf. These could be input to your program. Which of the methods you chose is more accurate? Can you think of a reason why it might be important to know the area of a leaf? Include any sketches of the leaf with your computations as well as the leaf taped to a paper. Be sure your methodology is clearly explained.

12. Find a map of an interesting place that comes with a scale and has distinct borders, for example suppose we pick Italy. By searching on google for 'outline map Italy' you'll find several. Do not pick a simple rectangular place, like the state of Pennsylvania. Find the area of that region using two different composite numerical integration techniques of your choice.

    One approach to this would be to print the map, align Italy on a coordinate system and measure different points that could be input to your code. Use at least 10 subintervals but if the geometry is interesting you may need to use more–explain your choice. Using your scale, how would you scale up your computations and compare to the real area? Be sure your methodology is clearly explained.

13. You are asked to come up with a concept for a new roller coaster at the Flux Theme Park. Your coaster is to be made of piecewise continuous functions that satisfy certain design constraints. Once the coaster concept is developed you will estimate the construction cost based on the assumption that the materials and labor required to build the coaster are proportional to the length of the track and the amount of support required.

    In addition to being a continuous track, your coaster must satisfy the following design requirements:

    (a) The first hill will be a parabola.
    (b) Your coaster design should include at least one addition hill segment that is a trig function. This can be of the form $A \cos(Bx + C) + D$ or $A \sin(Bx + C) + D$, where A, B, C, and D are constants.
    (c) There should be one clothoid loop. A clothoid loop has a tear drop shape and can be formed by joining 3 overlapping circles.

(d) You may need other functions to connect the track pieces–these can be any-
    thing you choose–but you should not have any obvious SHARP transitions.
    Your coaster must also satisfy the following physical constraints:
(e) The first hill needs to be the tallest.
(f) The first hill should be at least 50'

Calculating the cost: The total construction cost assumes that the cost is given
by $C = k_1 A + k_2 L$, where k_1 and k_2 are cost coefficients, A is the area below
the entire track and L is the length of the entire track. Use $k_1 = \$750/\text{ft}^2$ and
$k2 = \$185/\text{ft}$.

Area Below Track: Think about which parts of your coaster need support–for
example, you do not need support for the interior of your loop. You will need to
use approximate integration to calculate the area below your coaster (use Simpson's rule) and a sufficient number of subintervals. Comment on the accuracy of
your calculations and your choice of subintervals.

Length of Track: You should use the arc length formula described in below. Most
arc length problems result in integrals which cannot be evaluated exactly, with
the integral given by

$$L = \int_a^b \sqrt{1 + (f'(x))^2}\, dx$$

For this you will need to use numerical integration to approximate the resulting
integral. Use Simpson's rule with a sufficient number of subintervals.

Submit a detailed report on the coaster design to the CEO of Flux. Clear and
concise technical writing is important. Please take the time to create a report that
you are proud of.

3.5 Practical Numerical Integration

In practical applications, numerical integration may be embedded within a larger
program used to simulate some real-world phenomena. Is may be essentially that
the result has a specified accuracy or there could be consequences or even failure
in some other aspect of the underlying model. In this section, we address how to
implement numerical integration schemes to achieve a certain accuracy.

So the objective now is to compute

$$I = \int_a^b f(x)\, dx$$

with an error smaller than some prescribed tolerance ε. The methods presented here
are based on the trapezoid and Simpson's rules, but can easily be extended to other
methods.

Recall the error in the composite Simpson's rule estimate of I from Eq. (3.28)

$$I - S_N = -\frac{(b-a)h^4}{180}f^{(4)}(\xi)$$

where ξ is some (unknown) mean value point.

This formula is the basis for an approach to practical numerical integration using the following steps:

1. Find a bound M_4 for $\left|f^{(4)}(x)\right|$ on $[a, b]$.
2. Choose N such that

   (a) N is even, and
   (b) with $h = (b-a)/N$, this gives

$$\left|\frac{(b-a)h^4}{180}M_4\right| < \varepsilon \tag{3.29}$$

3. Evaluate S_N.

The first step is (typically) a paper and pencil derivation.

In applying (3.29), we can replace h by $(b-a)/N$, so that the goal is to find N so that

$$\left|\frac{(b-a)^5}{180N^4}M_4\right| < \varepsilon.$$

This gives

$$N > |b-a|\sqrt[4]{\frac{|b-a|M_4}{180\varepsilon}} \tag{3.30}$$

The following example demonstrates this process.

Example 14 Use the composite Simpson's rule to evaluate

$$I = \int_{-1}^{1}\frac{1}{x+2}dx$$

with an error smaller than 10^{-8}.

We saw in Example 12 in the previous section that $\left|f^{(4)}(x)\right| \le 24$ on $[-1, 1]$ and so we may take $M_4 = 24$. It follows from (3.30) that we desire

$$N > 2\left[\frac{2 \cdot 24}{180 \times 10^{-8}}\right]^{1/4} \approx 143.7$$

N must be an even integer, and so we choose $N = 144$.

The following Python command yields the result shown.

```
>>> S = simpsum(lambda x: 1 / (x + 2), -1, 1, 144)
>>> print(S)
1.0986122898925912
```

The true value is $\ln 3 = 1.0986122886681098$ so that the actual error is only about 1×10^{-9} which is indeed smaller than the required tolerance.

Often, actual error turns out to be smaller than the predicted error since the bound on the fourth derivative plays a role. In particular, M_4 may be much greater than the value of $\left| f^{(4)}(\xi) \right|$ in (3.28). Still, this process is overall, a reliable and practical approach for implementing numerical integration. Moreover, the entire process can be automated except for the evaluation of M_4.

The following script implements all of this procedure once a value of M_4 is known. It simply computes N using (3.30) and then calls the previously defined simpsum function.

Program Python function for computing $\int_a^b f(x)\,dx$ with error less than eps using Simpson's rule

```
import numpy as np
from prg_simpsum import simpsum

def simpson(fcn, a, b, M4, eps):
    """
    Approximate the integral of the function
    'fcn' over the interval [a, b] to accuracy
    'eps'. 'M4' must be a bound on the fourth
    derivative of 'fcn' on the interval.

    """

    L = abs(b - a)
    N = int(np.ceil(L * np.sqrt(np.sqrt(L * M4 /
        180 / eps))))
    # N must be even
    if N % 2 == 1:
        N += 1

    return simpsum(fcn, a, b, N)
```

Note the use of abs(b-a). This allows the possibility that $b < a$ which is mathematically well-defined, and which is also handled perfectly well by Simpson's rule. If $b < a$, then the steplength h will be negative and the appropriately signed result will be obtained.

The use of this function for the integral of Example 14 is:

```
S = simpson(lambda x: 1/(x + 2), −1, 1, 24, 1e−8)
```

which gives the same result as before.

As we have pointed out repeatedly–in practice, things are often not as straightforward. In the case that the problem is defined only in terms of experimental data for example–how would a fourth derivative be computed? Or it may be complicated to even compute a fourth derivative much less determine a bound. In that case there is a simple alternative to the procedure described above. The idea is to start with a small number of subdivisions of the range of integration and continually double this until we get subsequent results which agree to within our tolerance. The error formula (3.28) can be used to help justify this.

Consider (3.28) for Simpson's rule using first N and then $2N$ intervals. Let $h = (b - a)/2N$, and then

$$E\,(S_N) = -\frac{(b - a)\,(2h)^4}{180} f^{(4)}\,(\xi_1) \tag{3.31}$$

$$E\,(S_{2N}) = -\frac{(b - a)\,h^4}{180} f^{(4)}\,(\xi_2) \approx \frac{E\,(S_N)}{16} \tag{3.32}$$

if $f^{(4)}$ is approximately constant over $[a, b]$. If $|S_N - S_{2N}| < \varepsilon$, and $16\,(I - S_{2N}) \approx (I - S_N)$, then it follows that

$$|I - S_{2N}| \approx \varepsilon/15 \tag{3.33}$$

so that second of these estimates is expected to be well within the prescribed tolerance.

Example 15 Recompute the integral $I = \int_{-1}^{1} \frac{1}{x+2} dx$ starting with $N = 2$ and repeatedly doubling N until two Simpson's rule estimates agree to within 10^{-10}.

The while loop below produces the results shown:

```
N = 2
new_S = simpsum(lambda x: 1/(x + 2), −1, 1, N)
old_S = 0
while abs(old_S − new_S) > 1e−10:
    old_S = new_S
    N *= 2
    new_S = simpsum(lambda x: 1/(x + 2), −1, 1, N)
    print(N, new_S)
```

4	1.1000000000
8	1.0987253487
16	1.0986200427
32	1.0986127864
64	1.0986123200
128	1.0986122906
256	1.0986122888
512	1.0986122887
1024	1.0986122887

which is essentially the same answer as we obtained previously.

Note that the computation of Example 15 required about twice as many points in total since the earlier function evaluations are repeated. It is possible to remove this repetition by making efficient use of the midpoint and trapezoid rules to construct Simpson's rule as in (3.27) so that only the new nodes are evaluated at each iteration. We shall not concern ourselves with this aspect here. It is possible to construct an automatic integration script based on this repeated doubling of the number of intervals (see exercises!).

In taking a closer look, we see that the comparison of the errors in two successive Simpson's rule estimates, Eqs. (3.31) and (3.32) of an integral can be used further to obtain greater accuracy at virtually no greater computational cost. The approximate equation

$$16 \left(I - S_{2N} \right) \approx \left(I - S_N \right)$$

can be rearranged to give

$$I \approx \frac{16 S_{2N} - S_N}{15} \tag{3.34}$$

and this process should eliminate the most significant contribution to the error.

Using $N = 2$ in (3.34) and with $h = (b - a) / 2N$, we get

$$15I \approx \frac{16h}{3} \{ f (a) + 4f (a + h) + 2f (a + 2h) + 4f (a + 3h) + f (b) \}$$
$$- \frac{2h}{3} \{ f (a) + 4f (a + 2h) + f (b) \}$$

which simplifies to yield

$$I \approx \frac{2h}{45} [7f (a) + 32f (a + h) + 12f (a + 2h) + 32f (a + 3h) + 7f (b)] \tag{3.35}$$

This formula is in fact the interpolatory quadrature rule using five equally spaced nodes which actually could have been derived just like we did for the other interpolatory rules. The error in using Eq. (3.35) has the form $C (b - a) h^6 f^{(6)} (\xi)$. The above process of cutting the steplength in half, and then removing the most significant error contribution could be repeated using Eq. (3.35) to get improved accuracy. This process is the underlying ideas behind *Romberg integration*.

In practice, Romberg integration is typically started from the trapezoid rule rather than Simpson's rule. In that case, Simpson's rule is actually rediscovered as the result of eliminating the second order error term between T_1 and T_2. You will see that

$$S_{2N} = \frac{4T_{2N} - T_N}{3} \tag{3.36}$$

(The verification of this result is left as an exercise.)

Romberg integration can be expressed more conveniently by introducing a notation for the various estimates obtained and arranging them as elements of a lower triangular matrix. This is done as follows. The first column has entries $R_{n,1}$ which are the trapezoid rule estimates using 2^{n-1} subdivisions:

$$R_{n,1} = T_{2^{n-1}}$$

The second column has the results that would be obtained using Simpson's rule, so that, for $n \geq 2$, we have

$$R_{n,2} = S_{2^{n-1}} = \frac{4T_{2^{n-1}} - T_{2^{n-2}}}{3} = \frac{4R_{n,1} - R_{n-1,1}}{3}$$

Now applying the improved estimate (3.34), or equivalently, (3.35), this gives

$$R_{n,3} = \frac{16R_{n,2} - R_{n-1,2}}{15}$$

for $n \geq 3$.

Continuing in this manner produces the following array;

$$
\begin{array}{llllll}
R_{1,1} \\
R_{2,1} & R_{2,2} \\
R_{3,1} & R_{3,2} & R_{3,3} \\
R_{4,1} & R_{4,2} & R_{4,3} & R_{4,4} \\
R_{5,1} & R_{5,2} & R_{5,3} & R_{5,4} & R_{5,5} \\
\vdots & \vdots & \vdots & \vdots & \vdots & \ddots \\
\vdots & \vdots & \vdots & \vdots & \vdots & \vdots & \ddots
\end{array}
$$

The arguments presented above can be extended with

$$R_{n,k+1} = \frac{4^k R_{n,k} - R_{n-1,k}}{4^k - 1} \tag{3.37}$$

to obtain entries further to the right in that table.

The following Python code illustrates the implementation of this process to compute the array as far as $R_{M,M}$.

Program Python function for Romberg integration as far as the entry $R_{M,M}$ where M is some maximum level to be used.

```python
import numpy as np
from prg_midpoint import midsum

def romberg(fcn, a, b, ML):
    """
    Integrate 'fcn' using Romberg integration
    based on the midpoint rule. Returns the full
    Romberg array to maximum level 'ML'.

    """

    N = 1
    R = np.zeros((ML, ML))
    R[0, 0] = (b - a) * (fcn(a) + fcn(b)) / 2
    for L in range(1,ML):
        M = midsum(fcn, a, b, N)
        R[L, 0] = (R[L - 1, 0] + M) / 2
        for k in range(L):
            R[L, k + 1] = ((4**(k + 1) * R[L, k]
                          - R[L - 1, k]) /
                          (4**(k + 1) - 1))
        N *= 2
    return R
```

Note the use of the appropriate midpoint rule to update the trapezoid rule R(L,1). You can also verify that

$$T_{2N} = \frac{T_N + M_N}{2}$$

to allow for a more efficient computation of subsequent trapezoid rule estimates by ultimately avoiding the need to re-evaluate the function at nodes that have already been used. (See the exercises for a verification of this identity.)

Example 16 Use Romberg integration with 6 levels to estimate the integrals

a. $\int_{-1}^{1} \frac{1}{x+2} dx$, and

b. $\frac{1}{\sqrt{2\pi}} \int_{0}^{2} \exp\left(-x^2/2\right) dx$

The following Python commands can be used with the function romberg above to yield the results shown.

Integral a.

```
>>> romberg (lambda x: 1 / (x + 2), -1, 1, 6)
```

```
[[ 1.33333   0.        0.        0.        0.        0.       ]
 [ 1.16666   1.11111   0.        0.        0.        0.       ]
 [ 1.11666   1.1       1.09925   0.        0.        0.       ]
 [ 1.10321   1.09872   1.09864   1.09863   0.        0.       ]
 [ 1.09976   1.09862   1.09861   1.09861   1.09861   0.       ]
 [ 1.09890   1.09861   1.09861   1.09861   1.09861   1.09861 ]]
```

and integral b.

```
>>> romberg (lambda x: 1 / np.sqrt (2 * np.pi)
               * np.exp(-x**2 / 2), 0, 2, 6)}
```

```
[[ 0.45293   0.        0.        0.        0.        0.       ]
 [ 0.46843   0.47360   0.        0.        0.        0.       ]
 [ 0.47501   0.47720   0.47744   0.        0.        0.       ]
 [ 0.47668   0.47724   0.47725   0.47724   0.        0.       ]
 [ 0.47710   0.47724   0.47724   0.47724   0.47724   0.       ]
 [ 0.47721   0.47724   0.47724   0.47724   0.47724   0.47724 ]]
```

The zeros represent the fact that the Python code uses a matrix to store R and these entries are never assigned values. In the mathematical algorithm these entries simply do not exist.

In practice, Romberg integration is used iteratively until two successive values agree to within a specified tolerance as opposed to performed for a fixed number of steps. Usually this agreement is sought among entries on the diagonal of the array, or between the values $R_{n,n-1}$ and $R_{n,n}$. The subroutine above can be modified for this (see exercises).

In practice, you may only have data values to work with, implying you have no control over the length of the subintervals or ultimately, the accuracy. On the other hand, if you have a function to evaluate, we have shown there are ways to choose the number of subintervals so that a desired accuracy can be obtained. Here we present one more approach called *adaptive quadrature*. Adaptive integration algorithms use the basic principle of subdividing the range of integration but without the earlier insistence on using uniform subdivisions throughout the range. We present these ideas using Simpson's rule for the integration scheme in order to relay the key points however, any basic rule, or indeed Romberg integration, could be applied.

The two primary forms of adaptive quadrature are

1. Decide on an initial subdivision of the interval into N subintervals $\left[x_k, x_{k+1}\right]$ with

$$a = x_0 < x_1 < x_2 < \cdots < x_N = b$$

and then us the composite Simpson's rule with continual doubling of the number of intervals in each of the N original subintervals in turn. (Typical choices of the initial subdivision are five or 20 equal subintervals.)

2. Start with the full interval and apply Simpson's rule to the current interval and to each of its two halves. If the results agree to within the required accuracy, then the partial result is accepted, and the algorithm moves to the next subinterval. If the estimates do not agree to the required accuracy, then continue working with one half until accuracy is obtained, then move to the next half.

We demonstrate the first approach and leave the second as an exercise.

Example 17 Compute

$$\ln(3) = \int_{-1}^{1} \frac{1}{x+2} dx$$

with error less than 10^{-6} using adaptive Simpson's rule with $N = 5$ initial subdivisions.

On each of the intervals $[-1, -3/5]$, $[-3/5, 1/5]$, $[-1/5, 1/5]$, $[1/5, 3/5]$, and $[3/5, 1]$ we use continual halving of the steplengths until the Simpson's rule estimates agree to within $10^{-6}/5$.

On $[-1, -3/5]$ we get the results:
$N = 2$ $S_2 = 0.336507936507936$
$N = 4$ $S_4 = 0.336474636474636$
$N = 8$ $S_8 = 0.336472389660795$
$N = 16$ $S_{16} = 0.336472246235863$

On the remaining intervals, the converged values are:
$[-3/5, -1/5]$ $S_{16} = 0.251314430427972$
$[-1/5, 1/5]$ $S_8 = 0.200670706389129$
$[1/5, 3/5]$ $S_8 = 0.167054088990710$
$[3/5, 1]$ $S_8 = 0.143100845625304$

and summing these, we obtain the approximation $\ln(3) \approx 1.098612317668978$ which has an error of about 2.9×10^{-8}.

If there were no repetitions of any function evaluations, this would have used $17 + 16 + 8 + 8 + 8 = 57$ points in all.

The actual error is much smaller than the required tolerance. One reason for this is that, as we saw in our discussion of Romberg integration, if $|S_N - S_{2N}| \approx \varepsilon$, then we expect $|I - S_{2N}| \approx \varepsilon/15$. The local tolerance requirement in Example 17 is therefore much smaller than is needed for the overall accuracy.

Efficient programming of adaptive quadrature is more complicated than the other schemes discussed here. It has significant advantages when the integrand varies more than in this example, but for relatively well-behaved functions such as we have encountered here, the more easily programmed Romberg technique or the use of an initial subdivision, as in Example 17 is probably to be preferred.

Exercises

1. Show that

$$S_2 = \frac{4T_2 - T_1}{3}$$

and verify the result for $\int_1^2 1/(x+2)\,dx$.

2. Use the result of the previous exercise to prove Eq. (3.36):

$$S_{2N} = \frac{4T_{2N} - T_N}{3}$$

3. Show that $T_2 = (T_1 + M_1)/2$ and therefore that

$$T_{2N} = \frac{T_N + M_N}{2}$$

4. Use Romberg integration with four levels to estimate $\int_0^2 \frac{4}{1+x^2}\,dx$.

5. Write a program to perform Romberg integration iteratively until two diagonal entries $R_{n-1,n-1}$ and $R_{n,n}$ agree to within a specified accuracy. Use your program to estimate $\int_{-1}^1 \frac{1}{1+x^2}\,dx$ with a tolerance of 10^{-8}.

6. Write a program which applies composite Simpson's rule with repeated doubling of the number of intervals in each of five initial subintervals. Verify the results of Example 15.

7. Use adaptive Simpson's rule quadrature with the interval being continually split to evaluate the integral of Example 17. How do the number of function evaluations compare?

8. Modify program for Exercise 6 to perform adaptive integration using composite Simpson's rule on each interval of a specified uniform partition of the original interval into N pieces. Use it to evaluate the integral of Example 17 with $N = 5$, 10 and 20 initial subintervals.

9. The following question was the 2017 Society for Industrial and Applied Mathematics M3 Challenge in modeling. See

https://m3challenge.siam.org/archives/2017

for more details and access to some related data.

From Sea to Shining Sea: Looking Ahead with the Unites States National Park Service The National Park System of the United States comprises 417 official units covering more than 84 million acres. The 100-year old U.S. National Park Service (NPS) is the federal bureau within the Department of the Interior responsible for managing, protecting, and maintaining all units within the National Park system, including national parks, monuments, seashores, and other historical sites.

Global change factors such as climate are likely to affect both park resources and visitor experience. and, as a result, the NPS's mission to "preserve unimpaired the natural and cultural resources and values of the National Park System for the enjoyment, education, and inspiration of this and future generations." Your team can provide insight and help strategize with the NPS as it starts its second century of stewardship of United States' park system.

(a) **Tides of Change**: Build a mathematical model to determine a sea level change risk rating of high, medium, or low for each of the five parks below for the next 10, 20, and 50 years.
Acadia National Park, Maine
Cape Hatteras National Seashore, North Carolina
Kenai Fjords National Park, Alaska
Olympic National Park, Washington
Padre Island National Seashore, Texas
You will want to gig for data on sea level to build the model. Explain your interpretation of high, medium, and low. Could your model realistically predict those levels for the next 100 years?

(b) **The Coast is Clear**? In addition to the phenomena listed above, the NPS is investigating the effects of all climate-related events on coastal park units. Develop a mathematical model that is capable of assigning a single climate vulnerability score to any NPS coastal unit. Your model should take into account both the likelihood and severity of climate-related events occurring in the park within the next 50 years. Some or all of the provided data may be used to assign scores to the five national park units identified in Question 1.

(c) **Let Nature take its Course** NPS works to achieve its mission with limited financial resources that may vary from year to year. In the event that costs such as those caused by climate-related events exceed revenues and funding, NPS must prioritize where to spend monies.
Consider incorporating visitor statistics and your vulnerability scores (and possibly other variables that may be considered priorities) to create a new model that predicts long-term changes in visitors for each park. Use this output to advise NPS where future financial resources should go.

3.6 Conclusions and Connections: Numerical Calculus

What have you learned in this chapter? And where does it lead?

In some respects the methods of numerical differentiation are similar to those of numerical integration, in that they are typically based on using (in this case, differentiating) an interpolation polynomial. We study interpolation explicitly in Chap. 6 and have avoided specific use of interpolation in this chapter.

There is, however, one major and important difference between numerical approaches to integration and differentiation. We saw in the previous sections that integration is numerically a highly satisfactory operation with results of high accuracy being obtainable in economical ways. This is because integration tends to smooth out the errors of the polynomial approximations to the integrand. Unfortunately the reliability and stability we find in numerical integration is certainly not reflected for differentiation which tends to exaggerate the error in the original polynomial approximation. Figure 3.3 illustrates this clearly.

It is worth noting that although numerical differentiation is likely to be unreliable (especially if high accuracy is needed), its symbolic counterpart is generally fairly easy. Computer Algebra Systems, such as Maple, need only program a small number of basic rules in order to differentiate a wide variety of functions. This is in direct contrast to the situation for integration. We have seen that numerical integration is reasonably easy, but symbolic integration is very difficult to program well. This is partly due to the fact that many elementary functions have no antiderivative that can be written in terms of the same set of elementary functions. Even where this is possible, there are no simple algorithms to determine which integration techniques should be applied. This contrast illustrates at a simple level the need for mathematicians, scientists and engineers to develop a familiarity with both numerical and symbolic computation. The blending of the two in solving more difficult problems remains a very active field of scientific research which is blurring the lines among science, computer science, and pure and applied mathematics.

For the reasons cited numerical differentiation is best avoided whenever possible. We will see in subsequent sections discussing Newton's method and the secant method, that if the true value of the derivative is of secondary importance, then a simple approximation to the derivative can be satisfactory. In Chap. 7, we consider the numerical solution of differential equations. The methods there are philosophically based on approximations to derivatives but we will see that relatively simple approaches can be powerful. This is due, at least in part, to the fact that solving a differential equation was historically referred to as integrating said equation. Some of the stability of numerical integration carries over to that setting.

What we have done on integration is only the beginning. Students of calculus are well aware that multivariable integration is often much harder than for a function of a single variable. That is certainly also true for numerical integration. Singular and infinite integrals create difficulties of their own, as do integrals of oscillatory functions. There is still active research on obtaining or improving accurate approximate integration "rules" for all these situations and for obtaining high degree polynomial precision in different circumstances. These include Gaussian quadrature, and variations on that theme.

3.7 Python Functions for Numerical Calculus

Python (or rather NumPy and SciPy) provide functions for some of the operations we have discussed in this chapter.

numpy.diff can be helpful in estimating derivatives. It computes differences of the entries of a NumPy array. If x is a NumPy array (vector) then:

```
>>> dx = np.diff(x)
```

generates the vector of differences

$$dx = [x[2] - x[1], x[3] - x[2], \ldots, x[N] - x[N - 1]]]$$

If we have a vector y of corresponding function values, the division np.diff(y) / np.diff(x) results in the vector of divided differences. This is the simplest approximations to the first derivative.

scipy.integrate.quad is a function from the SciPy package for basic numerical integration. Calling the function is fairly straightforward:

```
>>> from scipy.integrate import quad
>>> integral = quad(np.square, 0, 1)
```

This gives the result

```
(0.33333333333333337, 3.700743415417189e−15)
```

where the first entry in the output tuple is the result, and the second entry is the estimated absolute error in the result. Here np.square is the function (simply $f(x) = x^2$) being integrated over the interval [0, 1]. There are several optional parameters allowing more detailed configuration of the algorithm, see the documentation for SciPy.

The scipy.integrate module contains several other numerical integration functions using various methods. We encourage you to explore the SciPy documentation for more details.

Linear Equations

4

4.1 Introduction

The notion of solving systems of linear equations was likely introduced to you in secondary school. For example, you may have encountered problems like this; *All 200 students in your math class went on vacation. Some students rode in vans which hold 8 students each and some students rode in buses which hold 25 students each. How many of each type of vehicle did they use if there were 15 vehicles total?* If x is the number of vans and y is the number of buses we have the following two equations and two unknowns;

$$8x + 25y = 200$$
$$x + y = 15,$$

which we could express in matrix form as

$$\begin{bmatrix} 8 & 25 \\ 1 & 1 \end{bmatrix} \begin{bmatrix} x \\ y \end{bmatrix} = \begin{bmatrix} 200 \\ 15 \end{bmatrix}.$$

Although this example is trivial, it demonstrates the simple notion of going from a real world situation, to a linear model, to a matrix representation. In cases like this where the number of equations and the number of unknowns are both small, solutions can be found by hand using the familiar idea of eliminating the unknowns one-by-one by solving one equation for one of them in terms of the others. Eventually, we are left with just one equation in one unknown. Its solution can then be substituted back into the previous equation to find the next unknown and so on. This is the process of *Gauss elimination* which is presented in the next section. When it comes to real-world problems, the number of unknowns can be prohibitively large and the need for efficient (and accurate) algorithms is even to this day an active area of research.

© Springer International Publishing AG, part of Springer Nature 2018
P. R. Turner et al., *Applied Scientific Computing*, Texts in Computer Science,
https://doi.org/10.1007/978-3-319-89575-8_4

There are numerous situations that require the solution to a linear system of equations. We already saw this in the context of numerical integration when using quadrature schemes. We will see this later in the study of interpolation, whether by polynomials or by splines, and again in some finite difference methods for approximate derivatives. In the cases of cubic spline interpolation and finite difference methods, the resulting matrix of coefficients is tridiagonal and we will see that algorithms can be designed to exploit matrix structure and improve efficiency.

In this section we give an introduction to a deep topic and consider the solution of a square system of equations with a single right-hand side. These systems are expressed as

$$A\mathbf{x} = \mathbf{b} \tag{4.1}$$

where $\mathbf{x}, \mathbf{b}$ are n-vectors and A is an $n \times n$ real matrix. In full, this system would be written as

$$
\begin{bmatrix}
a_{11} & a_{12} & a_{13} & \cdots & a_{1n} \\
a_{21} & a_{22} & a_{23} & \cdots & a_{2n} \\
a_{31} & a_{32} & a_{33} & \cdots & a_{3n} \\
\vdots & \vdots & \vdots & \vdots & \vdots \\
a_{n1} & a_{n2} & a_{n3} & \cdots & a_{nn}
\end{bmatrix}
\begin{bmatrix}
x_1 \\
x_2 \\
x_3 \\
\vdots \\
x_n
\end{bmatrix}
=
\begin{bmatrix}
b_1 \\
b_2 \\
b_3 \\
\vdots \\
b_n
\end{bmatrix}
\tag{4.2}
$$

or even, writing each equation fully,

$$
\begin{aligned}
a_{11}x_1 + a_{12}x_2 + a_{13}x_3 + \cdots + a_{1n}x_n &= b_1 \\
a_{21}x_1 + a_{22}x_2 + a_{23}x_3 + \cdots + a_{2n}x_n &= b_2 \\
a_{31}x_1 + a_{32}x_2 + a_{33}x_3 + \cdots + a_{3n}x_n &= b_3 \\
\vdots \qquad \vdots \qquad \vdots \qquad \vdots \qquad \vdots \qquad \vdots \\
a_{n1}x_1 + a_{n2}x_2 + a_{n3}x_3 + \cdots + a_{nn}x_n &= b_n
\end{aligned}
\tag{4.3}
$$

Example 1 Suppose you are asked to design the first ascent and drop of a roller coaster. By studying other rides, you decide that the slope of the ascent should be 0.8 and the slope of the drop should be -1.6. You decide to connect these two line segments $L_1(x)$ and $L_2(x)$ with a quadratic function, $f(x) = ax^2 + bx + c$. Here x is measured in feet. For the track to be smooth there cannot be any abrupt changes so you want the linear segments to be tangent to the quadratic part at the transition points P and Q which are 100 feet apart in the horizontal direction. To simplify things, you place P at the origin. Write the equations that will ensure your roller coaster has smooth transitions and express this in matrix form.

A picture often helps. Consider Fig. 4.1. We need to find a, b, c, and d so that the continuity conditions hold, that is $f(0) = L_1(0)$ and $f(100) = L_2(100)$. Then, the smoothness conditions require $f'(0) = L_1'(0)$ and $f'(100) = L_2'(100)$. This gives

$$a(0^2) + b(0) + c = 0$$

Fig. 4.1 Conceptual sketch of the roller coaster design, linking two lines to a quadratic function

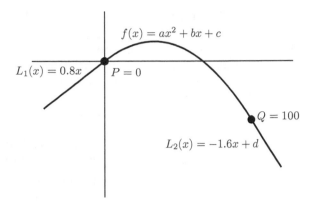

$$a(100^2) + b(100) + c = -1.6(100) + d$$

$$2a(0) + b = 0.8$$

$$2a(100) + b = -1.6$$

Rearranging terms we can express these four equations in matrix form as

$$\begin{bmatrix} 0 & 0 & 1 & 0 \\ 10^4 & 10^2 & 1 & -1 \\ 0 & 1 & 0 & 0 \\ 200 & 1 & 0 & 0 \end{bmatrix} \begin{bmatrix} a \\ b \\ c \\ d \end{bmatrix} = \begin{bmatrix} 0 \\ -1600 \\ 0.8 \\ -1.6 \end{bmatrix}$$

Note, early in the model development phase, it was easy to see that $c = 0$ (from the first continuity condition) and $b = 0.8$ (from the first smoothness condition). From there it is straightforward to substitute these values and solve for $a = -0.012$ and $d = 120$.

Another important problem of linear algebra arising frequently in applications is the *eigenvalue* problem, which requires the solution of the equation

$$\mathbf{Ax} = \lambda \mathbf{x} \tag{4.4}$$

for the *eigenvalues* λ and their associated *eigenvectors* $\mathbf{x}$. There are many techniques available for this problem and we introduce one iterative method later as well as an extension using our foundational linear equations techniques.

The ideas presented here are foundational and many variations arise. In practice, the same linear system may need to be solved but with multiple right-hand sides. That situation can be handled using the same methods presented in the next few sections with only slight modifications.

4.2 Gauss Elimination

The idea of Gaussian elimination is not new to you (think about the simple van-bus example above). However, the approach taken now is to streamline the procedure as a systematic process that can be written as an algorithm. Then, we discuss performance issues and improvements.

The first step in the Gauss elimination process is to eliminate the unknown x_1 from every equation in (4.3) except the first. The way this is achieved is to subtract from each subsequent equation an appropriate multiple of the first equation. Since the coefficient of x_1 in the first equation is a_{11} and that in the jth equation is a_{j1}, it follows that subtracting $m = \dfrac{a_{j1}}{a_{11}}$ times the first equation from the jth one will result in the coefficient of x_1 giving

$$a_{j1} - ma_{11} = a_{j1} - \frac{a_{j1}}{a_{11}}a_{11} = 0$$

The new system of equations is

$$\begin{aligned}
a_{11}x_1 + a_{12}x_2 + a_{13}x_3 + \cdots + a_{1n}x_n &= b_1 \\
a'_{22}x_2 + a'_{23}x_3 + \cdots + a'_{2n}x_n &= b'_2 \\
a'_{32}x_2 + a'_{33}x_3 + \cdots + a'_{3n}x_n &= b'_3 \\
\cdots \quad \cdots \quad \cdots \quad \cdots &= \cdots \\
a'_{n2}x_2 + a'_{n3}x_3 + \cdots + a'_{nn}x_n &= b'_n
\end{aligned} \tag{4.5}$$

or, in matrix form,

$$\begin{bmatrix}
a_{11} & a_{12} & a_{13} & \cdots & a_{1n} \\
0 & a'_{22} & a'_{23} & \cdots & a'_{2n} \\
0 & a'_{32} & a'_{33} & \cdots & a'_{3n} \\
\vdots & \vdots & \vdots & \vdots & \vdots \\
0 & a'_{n2} & a'_{n3} & \cdots & a'_{nn}
\end{bmatrix}
\begin{bmatrix}
x_1 \\ x_2 \\ x_3 \\ \vdots \\ x_n
\end{bmatrix}
=
\begin{bmatrix}
b_1 \\ b'_2 \\ b'_3 \\ \vdots \\ b'_n
\end{bmatrix} \tag{4.6}$$

where the modified coefficients are given by

$$a'_{jk} = a_{jk} - \frac{a_{j1}}{a_{11}}a_{1k}$$

$$b'_j = b_j - \frac{a_{j1}}{a_{11}}b_1 \tag{4.7}$$

This particular operation can be described by the loop
```
for j=2:n
```
$$m := \frac{a_{j1}}{a_{11}}; \; a_{j1} := 0$$
```
    for k=2:n
```
$$a'_{jk} := a_{jk} - ma_{1k}$$

end

$$b'_j := b_j - mb_1$$

end

In the general algorithm, and in all that follows, the $'$ notation will be dropped and elements of the matrix and components of the right-hand side will just be overwritten by their new values. With this convention the modified rows of (4.6) represent the system

$$
\begin{bmatrix}
a_{22} & a_{23} & \cdots & a_{2n} \\
a_{32} & a_{33} & \cdots & a_{3n} \\
\vdots & \vdots & \vdots & \vdots \\
a_{n2} & a_{n3} & \cdots & a_{nn}
\end{bmatrix}
\begin{bmatrix}
x_2 \\
x_3 \\
\vdots \\
x_n
\end{bmatrix}
=
\begin{bmatrix}
b_2 \\
b_3 \\
\vdots \\
b_n
\end{bmatrix}
\tag{4.8}
$$

The procedure can be repeated for (4.8) but next to eliminate x_2 and so on until what remains is a *triangular* system

$$
\begin{bmatrix}
a_{11} & a_{12} & a_{13} & \cdots & a_{1n} \\
0 & a_{22} & a_{23} & \cdots & a_{2n} \\
0 & 0 & a_{33} & \cdots & a_{3n} \\
\vdots & \vdots & \vdots & \ddots & \vdots \\
0 & 0 & 0 & \cdots & a_{nn}
\end{bmatrix}
\begin{bmatrix}
x_1 \\
x_2 \\
x_3 \\
\vdots \\
x_n
\end{bmatrix}
=
\begin{bmatrix}
b_1 \\
b_2 \\
b_3 \\
\vdots \\
b_n
\end{bmatrix}
\tag{4.9}
$$

This is called the "elimination" phase, or *forward elimination*, of Gauss elimination and can be expressed as an algorithm as follows.

Algorithm 3 *Forward elimination phase of Gauss elimination*

Input $n \times n$ matrix A, right-hand side n-vector **b**
Elimination
  for i=1:n-1
    for j=i+1:n
      $m := \dfrac{a_{ji}}{a_{ii}};\ a_{ji} := 0$
      for k=i+1:n
        $a_{jk} := a_{jk} - m a_{ik}$
      end
      $b_j := b_j - mb_i$
    end
  end
Output   triangular matrix A, modified right-hand side **b**

Example 2 Use Gauss elimination to solve the system

$$\begin{bmatrix} 1 & 3 & 5 \\ 3 & 5 & 5 \\ 5 & 5 & 5 \end{bmatrix} \begin{bmatrix} x \\ y \\ z \end{bmatrix} = \begin{bmatrix} 9 \\ 13 \\ 15 \end{bmatrix}$$

The first step of the elimination is to subtract multiples of the first row from each of the other rows. The appropriate multipliers to use for the second and third rows are $3/1 = 3$ and $5/1 = 5$ respectively. The resulting system is

$$\begin{bmatrix} 1 & 3 & 5 \\ 0 & -4 & -10 \\ 0 & -10 & -20 \end{bmatrix} \begin{bmatrix} x \\ y \\ z \end{bmatrix} = \begin{bmatrix} 9 \\ -14 \\ -30 \end{bmatrix}$$

Next $(-10)/(-4) = 5/2$ of the second row must be subtracted from the third one in order to eliminate y from the third equation:

$$\begin{bmatrix} 1 & 3 & 5 \\ 0 & -4 & -10 \\ 0 & 0 & 5 \end{bmatrix} \begin{bmatrix} x \\ y \\ z \end{bmatrix} = \begin{bmatrix} 9 \\ -14 \\ 5 \end{bmatrix}$$

This completes the forward elimination for this example.

The final row is the equation

$$5z = 5$$

so $z = 1$. Substituting this back into the second equation, we obtain

$$-4y - 10(1) = -14$$

and hence, $y = 1$. Finally substituting both these into the first equation, we get

$$x + 3(1) + 5(1) = 9$$

giving $x = 1$. The solution vector is $[1, 1, 1]^T$. (Throughout, the superscript T denotes the transpose of the vector. The original vector was a column, so the T is a row).

The second phase of the process used above is the *back substitution* phase of Gauss elimination. It can be summarized as follows.

Algorithm 4 *Back substitution phase of Gauss elimination*

Input $n \times n$ triangular matrix A, right-hand side n-vector **b**
    (typically the output from the forward elimination stage)
Elimination
    $x_n := b_n/a_{nn}$ (Solve final equation)
    for i=n-1:-1:1
      for j=i+1:n
        $b_i := b_i - a_{ij}x_j$ (Substitute known values and subtract)
      end
      $x_i := b_i/a_{ii}$
    end
Output   Solution vector **x**

By counting the number of multiplications and divisions in the two phases of the
Gauss elimination algorithms one can see that approximately $n^3/3$ multiplications
and $n^2/2$ divisions are needed. This may be reasonable if n is not too large, but in
practice if the linear system is the result of modeling a large-scale scenario then n
may be on the order of millions, meaning n^3 would be an exhausting number of
computations (and if you are thinking back to the chapter on floating point, errors,
you are not wrong in thinking there may be even more issues to consider!)

To this end, the whole Gauss elimination process is fairly easy to program and does
provide a powerful basic tool for solving linear systems. **But** it has its shortcomings
as the following simple example demonstrates.

Example 3 Solve the following system using Gauss elimination

$$\begin{bmatrix} 7 & -7 & 1 \\ -3 & 3 & 2 \\ 7 & 7 & -72 \end{bmatrix} \begin{bmatrix} x \\ y \\ z \end{bmatrix} = \begin{bmatrix} 1 \\ 2 \\ 7 \end{bmatrix}$$

Following the same algorithm as we used in the previous example, we begin by
subtracting $(-3)/7$ times the first row from the second and $7/7$ times the first row
from the third. The resulting *partitioned* matrix $\left[A \vdots \mathbf{b} \right]$ is

$$\begin{bmatrix} 7 & -7 & 1 & \vdots & 1 \\ 0 & 0 & 17/7 & \vdots & 17/7 \\ 0 & 14 & -8 & \vdots & 6 \end{bmatrix},$$

but at the next step we must subtract $14/0$ times the second row from the third!

By looking at the partitioned system above, the second row implies that $z = 1$ and then this could be used in the third equation to find y and so on. However, one must consider how to move forward within an algorithm. The issue in Example 3 could have been addressed simply by exchanging the second and third rows, to obtain the revised system

$$\begin{bmatrix} 7 & -7 & 1 & \vdots & 1 \\ 0 & 14 & -8 & \vdots & 6 \\ 0 & 0 & 17/7 & \vdots & 17/7 \end{bmatrix},$$

which is now a triangular system. The back substitution algorithm can now be used to complete the solution: $[x, y, z]^T = [1, 1, 1]^T$.

Example 3 demonstrates the need for *pivoting*, which is standard to avoid dividing by zero.

4.2.1 Pivoting in Gauss Elimination

At each step of the basic Gauss elimination algorithm the diagonal entry that is being used to generate the multipliers is known as the pivot element. In Example 3, we saw the effect of a zero pivot. The basic idea of *partial pivoting* is to avoid zero (or near-zero) pivot elements by performing row interchanges. The details will be discussed shortly, but first we revisit Example 3 to highlight the need to avoid near-zero pivots.

Example 4 Solve the system of Example 3 using Gauss elimination with four decimal place arithmetic.

At the first step, the multiplier used for the second row is $(-3)/7 = 0.4286$ to four decimals. The resulting system is then

$$\begin{bmatrix} 7 & -7 & 1 & \vdots & 1 \\ 0 & -0.0002 & 2.4286 & \vdots & 2.4286 \\ 0 & 14 & -8 & \vdots & 6 \end{bmatrix}$$

The multiplier used for the next step is then $14/(-0.0002) = -70000$ and the final entries are

$$\begin{bmatrix} 7 & -7 & 1 & \vdots & 1 \\ 0 & -0.0002 & 2.4286 & \vdots & 2.4286 \\ 0 & 0 & 169994 & \vdots & 170008 \end{bmatrix}$$

which gives the solutions (computed to four decimal places)

$$z = 170008/(169994) = 1.0001$$
$$y = (2.4286 - (2.4286)(1.0001))/(-0.0002) = 1.2143$$
$$x = (1 - ((-7)(1.2143) + (1)(1.0001)))/7 = 1.2143$$

Four decimal place arithmetic has left the results not even accurate to one decimal place.

The problem in Example 4 is that the second pivot entry is dominated by its roundoff error (in this case, it *is* its roundoff error). This results in an unfortunately large multiplier being used which magnifies the roundoff errors in the other entries to the point where the accuracy of the final result is affected.

The basic idea of partial pivoting is to search the *pivot column* to find its largest element (in absolute value) on or below the diagonal. The *pivot row* is then interchanged with this row so that this largest element becomes the pivot for this step.

Example 5 Repeat Example 4 using partial pivoting.

The first stage of elimination would still be

$$\begin{bmatrix} 7 & -7 & 1 & 1 \\ 0 & -0.0002 & 2.4286 & 2.4286 \\ 0 & 14 & -8 & 6 \end{bmatrix}.$$

Exchanging the 2nd and 3rd rows gives

$$\begin{bmatrix} 7 & -7 & 1 & 1 \\ 0 & 14 & -8 & 6 \\ 0 & -0.0002 & 2.4286 & 2.4286 \end{bmatrix}.$$

which has a multiplier of $(-0.0002/14) = 0$ to 4 decimal places giving a final array of

$$\begin{bmatrix} 7 & -7 & 1 & 1 \\ 0 & 14 & -8 & 6 \\ 0 & 0 & 2.4286 & 2.4286 \end{bmatrix}.$$

This yields the exact solution $[1, 1, 1]^T$.

The use of pivoting requires only a small change to the Gauss elimination algorithm.

Algorithm 5 *Forward elimination for Gauss elimination with partial pivoting*

> *Input* $n \times n$ matrix A, right-hand side n-vector **b**
> *Elimination*
>   for i=1:n-1
>   Find $p \geq i$ such that $|a_{pi}| = \max\left(|a_{ji}| : j \geq i\right)$
>   Interchange rows i and p
>   for j=i+1:n
> $m := \dfrac{a_{ji}}{a_{ii}}; \; a_{ji} := 0$
>   for k=i+1:n
> $a_{jk} := a_{jk} - m a_{ik}$
>   end
> $b_j := b_j - m b_i$
>   end
>   end
> *Output*   triangular matrix A, modified right-hand side **b**

NumPy's max and argmax functions make this easy to implement. Remember though that the position of the maximum will be given relative to the vector to which max/argmax is applied. Thus, for example, to find the largest entry in positions 0 through 9 of the vector $[1, 3, 5, 7, 9, 0, 2, 8, 6, 4]$ we can use

```
v = np.array([1, 3, 5, 7, 9, 0, 2, 8, 6, 4])
mx = np.max(v[5:])
pmax = np.argmax(v[5:])
```

to get the output

```
>>> mx
8

>>> pmax
2
```

Here 8 is indeed the maximum entry in the appropriate part of the vector $[0, 2, 8, 6, 4]$ and it is the third component of this vector. To get its position in the original vector we simply add 5 to account for the 5 elements that were omitted from our search.

Program   Python function for Gauss elimination with partial pivoting

```
def gepp(A, b):
    """

    Solve A x = b using Gauss elimination with
    partial pivoting for the square matix 'A' and
    vector 'b'.
```

```
    """

    # Avoid altering the passed-in arrays
    A = A.copy()
    b = b.copy()

    n = A.shape[0]   # n is number of rows of A
    # Forward elimination
    for i in range(n-1):
        # find the pivot position
        p = np.argmax(np.abs(A[i:,i]))
        p += i
        # row interchange
        A[[i, p], :] = A[[p, i], :]
        # interchange on right-hand side
        b[[i, p]] = b[[p, i]]
        for j in range(i + 1, n):
            m = A[j, i] / A[i, i]
            A[j, i] = 0
            A[j, i + 1:] -= m * A[i, i + 1:]
            b[j] -= m * b[i]
    # Back substitution
    x = np.zeros_like(b)
    x[-1] = b[-1] / A[-1, -1]
    for i in range(n-1,-1,-1):
        for j in range(i+1, n):
            b[i] -= A[i, j] * x[j]
        x[i] = b[i] / A[i, i]

    return x
```

Note that it is not strictly necessary to perform the row interchanges explicitly. A permutation vector could be stored which carries the information as to the order in which the rows were used as pivot rows. For our current purpose, the advantage of such an approach is outweighed by the additional detail that is needed. NumPy's vector operations make the explicit row interchanges easy and fast.

To test our algorithm, we shall use the *Hilbert matrix* which is defined for dimension n as

$$
H_n = \begin{bmatrix}
1 & 1/2 & 1/3 & \cdots & 1/n \\
1/2 & 1/3 & 1/4 & \cdots & 1/(n+1) \\
1/3 & 1/4 & 1/5 & \cdots & 1/(n+2) \\
\cdot & \cdot & \cdot & \cdots & \cdot \\
1/n & 1/(n+1) & 1/(n+2) & \cdots & 1/(2n-1)
\end{bmatrix}
$$

The elements are

$$h_{ij} = \frac{1}{i + j - 1}$$

For the right-hand side, we shall impose a simple test for which it is easy to recognize the correct solution. By taking the right-hand side as the vector of row-sums, we force the true solution to be a vector of ones. Note how NumPy's sum function is configured to sum across rows by the argument axis = 1.

Example 6 Solve the Hilbert matrix system of dimension 6 using Gauss elimination with partial pivoting.

The *n*-dimensional Hilbert matrix is available as a function in the SciPy package: hilbert(n). We can therefore use the following commands to generate the solution of ones:

```
from scipy.linalg import hilbert
import numpy as np
from prg_gepp import gepp

H = hilbert(6)
rsH = np.sum(H, axis = 1)
x = gepp(H, rsH)
np.set_printoptions(precision = 4)

>>> print(x)
[ 1.   1.   1.   1.   1.   1.]
```

At this level of accuracy, the results confirm that our algorithm is working. However if the solution is displayed again using higher precision we get

```
np.set_printoptions(precision = 15)

>>>print(x)
[ 0.999999999999072   1.000000000026699
  0.999999999818265   1.000000000474873
  0.999999999473962   1.000000000207872]
```

which suggests that there is considerable build-up of roundoff error in this solution.

This loss of accuracy is due to the fact that the Hilbert matrix is a well-known example of an *ill-conditioned* matrix. An ill-conditioned system is one where the accumulation of roundoff error can be severe and/or the solution is highly sensitive to small changes in the data. In the case of Hilbert matrices, the severity of the ill-conditioning increases rapidly with the dimension of the matrix.

For the corresponding 10×10 Hilbert matrix system to that used in Example 6, the "solution vector" obtained using the Python code above was

[1.0000, 1.0000, 1.0000, 1.0000, 0.9999, 1.0003, 0.9995, 1.0005, 0.9998, 1.0001]
showing significant errors – especially bearing in mind that Python is working internally with about 15 significant (decimal) figures.

Why do we believe these errors are due to this notion of ill-conditioning rather than something inherent in the program?

The NumPy random module's rand function can be used to generate random matrices. Using the same approach as above we can choose the right-hand side vector to force the exact solution to be a vector of ones. The computed solution can then be compared with the exact one to get information about the performance of the algorithm. The commands listed below were used to repeat this experiment with one hundred different matrices of each dimension from 4×4 up to 10×10.

```
np.random.seed(42)
E = np.zeros((100, 7))
for n in range(4,11):
    for k in range(100):
        A = np.random.uniform(-5, 5, size=(n, n))
        b = np.sum(A, axis = 1)
        x = gepp(A, b)
        E[k, n - 4] = np.max(np.abs(x - 1))
```

Each trial generates a random matrix with entries in the interval $[-5, 5]$. The matrix E contains the largest individual component-error for each trial. Its largest entry is therefore the worst single component error in the complete set of 700 trials:

```
>>> np.max(E)
4.99822405686e-13
```

This output shows that none of these randomly generated test examples generated errors that were quite large. Of course, some of these random matrices could be somewhat ill-conditioned themselves. The command

```
>>> np.sum(E > 1e-14)
43
```

counts the number of trials for which the worst error was greater than 10^{-14}. There were just 43 such out of the 700.

Note that NumPy's sum function sums all entries of a matrix when no axis argument is given. Therefore the first np.sum in the command above generates a count of the number of cases where $E > 10^{-14}$ across both dimensions. Gauss elimination with partial pivoting appears to be a highly successful technique for solving linear systems.

4.2.2 Tridiagonal Systems

We will revisit the heat equation example from Chap. 3 as an example of how tridiagonal matrices arise in practice. We will also encounter these in the context of cubic spline interpolation later. The tridiagonal algorithm presented here is just one example of how matrix structure can lead to improved efficiency.

A tridiagonal matrix has its only nonzero entries on its main diagonal and immediately above and below this diagonal. This implies the matrix can be stored simply by storing three vectors representing these diagonals. Let a generic tridiagonal system be denoted as

$$
\begin{bmatrix}
a_1 & b_1 & & & & \\
c_1 & a_2 & b_2 & & & \\
 & c_2 & a_3 & b_3 & & \\
 & & \ddots & \ddots & \ddots & \\
 & & & c_{n-2} & a_{n-1} & b_{n-1} \\
 & & & & c_{n-1} & a_n
\end{bmatrix}
\begin{bmatrix}
x_1 \\ x_2 \\ x_3 \\ \vdots \\ x_{n-1} \\ x_n
\end{bmatrix}
=
\begin{bmatrix}
r_1 \\ r_2 \\ r_3 \\ \vdots \\ r_{n-1} \\ r_n
\end{bmatrix}
\tag{4.10}
$$

where all the other entries in the matrix are zero.

At each step of the elimination, there is only one entry below the diagonal and so there is only one multiplier to compute. When a (multiple of a) row is subtracted from the one below it, the only positions affected are the diagonal and the immediate subdiagonal. The latter is of course going to be zeroed.

These observations imply that the inner loops of the Gauss elimination algorithm, Algorithm 3, collapse to single operations and a drastic reduction in computational effort.

Algorithm 6 *Gauss elimination for a tridiagonal system*

> *Input* $n \times n$ tridiagonal matrix A as in (4.10), right-hand side n-vector $\mathbf{r}$
> *Elimination*
>   for i=1:n−1
> $m := \dfrac{c_i}{a_i}; \; c_i := 0$
> $a_{i+1} := a_{i+1} - mb_i$
> $r_{i+1} := r_{i+1} - mr_i$
>   end
> *Back substitution*
> $x_n = r_n / a_n$
>   for i=n−1:−1:1
> $x_i = (r_i - b_i x_{i+1}) / a_i$
>   end
> *Output*   Solution vector $\mathbf{x}$

Example 7 Steady state heat equation in one dimension: Consider a simplified version of the heat model on a wire (Eq. 3.1) but set $u_t = 0$ (i.e. the temperature along the wire is no longer changing in time). This example demonstrates how to express $-Ku_{xx} = f(x)$ as a linear system. To see this, we will focus on the mathematics itself and not necessarily worry about the physical parameters. To this end, assume the temperature at the boundary is fixed at $0\,°C$ and $f(x) = 1\,°C$ for all the interior values of $[0, 1]$ and let $K = 1$. Using numerical derivatives, approximate the temperatures at five equally spaced points along the interior of the wire by setting up and solving a linear system.

Since we have 5 interior locations for $0 \leq x \leq 1$, $\Delta x = \frac{1}{6}$, and we can define the nodes $x_i = i\Delta x$ for $i = 0, 1, 2, \ldots 6$. As before, let u_i denote the approximation to $u(x_i)$. Our boundary requirement means that $u_0 = u_6 = 0$ and the unknowns are $u_1, u_2, \ldots, u_5$.

Using our previous approach for numerical derivatives the ith equation is given by

$$-\left[\frac{u_{i-1} - 2u_i + u_{i+1}}{\Delta x^2}\right] = f_i. \tag{4.11}$$

Note, we no longer need to the superscript k since this model is not time-dependent. Also note that the ith equation implies that the temperature at a location only depends on the temperatures immediately next to it. This will lead to a tridiagonal coefficient matrix of the form

$$\begin{bmatrix} 2 & -1 & & & \\ -1 & 2 & -1 & & \\ & -1 & 2 & -1 & \\ & & -1 & 2 & -1 \\ & & & -1 & 2 \end{bmatrix} \begin{bmatrix} u_1 \\ u_2 \\ u_3 \\ u_4 \\ u_5 \end{bmatrix} = \Delta x^2 \begin{bmatrix} 1 \\ 1 \\ 1 \\ 1 \\ 1 \end{bmatrix}$$

Using $\Delta x = 1/6$, the elimination phase results in the bidiagonal (augmented) matrix

$$\begin{bmatrix} 2 & -1 & & & & : & 0.1667 \\ 0 & 1.5 & -1 & & & : & 0.25 \\ & 0 & 1.3333 & -1 & & : & 0.3333 \\ & & 0 & 1.25 & -1 & : & 0.4167 \\ & & & 0 & 1.2 & : & 0.5 \end{bmatrix}$$

from which we obtain the solution

$$u_5 = 0.5/1.2 = 0.4167$$
$$u_4 = (0.4167 + 0.4167)/1.25 = 0.6667$$
$$u_3 = (0.3333 + 0.6667)/1.3333 = 0.7500$$
$$u_2 = (0.25 + 0.7500)/1.5 = 0.6667$$
$$u_1 = (0.1667 + 0.6667)/2 = 0.4167$$

Although plotting this solution is crude for this small number of points, you should consider doing so and then ask yourself why the shape of this temperature distribution makes sense.

The important difference between this algorithm and Algorithm 3 is that it requires a much smaller amount of computation. For a square system, we stated earlier that approximately $n^3/3$ multiplications and $n^2/2$ divisions are needed. For a tridiagonal system, using Algorithm 6 reduces these counts to just $3n$ and $2n$ respectively.

Exercises

1. Solve the system

$$
\begin{bmatrix}
1 & 2 & 3 & 4 \\
2 & 2 & 3 & 4 \\
3 & 3 & 3 & 4 \\
4 & 4 & 4 & 4
\end{bmatrix}
\begin{bmatrix}
w \\
x \\
y \\
z
\end{bmatrix}
=
\begin{bmatrix}
20 \\
22 \\
22 \\
24
\end{bmatrix}
$$

   using Gauss elimination without pivoting.
2. Repeat Exercise 1 using partial pivoting.
3. Write a program to solve a system of linear equations by Gauss elimination without pivoting. Test your program by re-solving Exercise 1.
4. Solve the Hilbert matrix systems (with exact solution vectors all ones) for dimensions $4, 5, \ldots, 10$ using Gauss elimination.
5. Repeat Exercise 4 using partial pivoting.
6. The roller coaster from Example 1 might look smooth but it won't feel smooth because the piecewise defined function does not have continuous second derivatives. You decide to improve the design by using a quadratic function, $Q(x) = ax^2 + bx + c$ only on the interval $[10, 90]$ and connecting it to the linear segments using two cubic polynomials as follows:

$$G(x) = kx^3 + lx^2 + mx + n, 0 \le x < 10$$

$$H(x) = px^3 + qx^2 + rx + s, 90 \le x < 100.$$

   (a) Write a new system of equations that ensures the coaster feels smooth at the transition points.
   (b) Express this new system in the form Au = w.
   (c) Solve the resulting system and provide a plot for the roller coaster.

7. Write a program to solve a tridiagonal system using Gauss elimination. Test it on the 50×50 system with any diagonal entries of your choice and a right hand side so that the exact solution vector is $[1, 1, 1, \ldots, 1]'$.
8. Consider the heat flow model in Example 7 but this time, consider 100 interior nodes and solve the problem again with $f(x) = 1$. Plot your solution and explain why the shape of the graph makes sense in terms of the equation you are solving.
9. Consider the heat flow model in Example 7 but this time, suppose that your true solution is $u^* = sin(\pi x)$.

(a) What is $f(x)$ given u^*?
(b) Write a subroutine that takes in a vector x and outputs the values of $f(x)$ and another that calculates $u^*(x.)$
(c) Solve this problem using $n = 10$ interior nodes and plot your computed solution along with the true solution on the same set of graph.
(d) Use your tridiagonal solver for this problem on a sequence of grids using $n = 25, 50, 100, 200$ interior nodes. Calculate the errors using $e_n = (1/\sqrt{(n)})$ $||u - u^*||_2$ for each of those discretizations. Note, here we scale the 2-norm error to account for the fact that the vectors are changing in length. Next, calculate the ratio of consecutive errors and explain what you see.

10. We now revisit the image distortion example from the numerical calculus chapter which was modeled with

$$u_t - Ku_{xx} - Ku_{yy} = f(x, y, t).$$

You (may now) know from the chapter on numerical derivatives that *explicit* methods can be highly unstable and care must be taken when choosing the time-step size and spatial discretization. *Implicit* methods can help overcome this but require a linear solve at each time step. The backward Euler approximation is one such approach and this will be covered in depth soon! For now, we can implement that method simply considering all the values on the right hand side evaluated at the time k + 1 instead (we will derive this as a numerical approach to differential equations later). This means we have unknowns on both sides of the equation and with some rearranging, we can arrive at a linear system of the form $A\mathbf{u}^{k+1} = \mathbf{b}$

(a) To check your understanding. Starting with Eq. (3.11), re-write the right hand side with all values at k+1 in time. Now re-arrange terms so that your unknowns are on the left and your knowns are on the right.
(b) How do you now express this as a linear system? You need to translate your unknowns on the spatial grid into an ordered vector so that you can determine the entries in the matrix A. For consistency, start at the bottom and move left to right in x and then up to the next y value, continuing left to right.
    For example at some fixed time on a 3 × 3 grid of unknowns is shown in Fig. 4.2.
    What does the resulting linear system look like for the 2-D diffusion equation at each time step for this 3 × 3 example above? What is b? Generalize this to any size grid. How has the matrix structure changed in comparison to the 1-D case?
(c) Outline, in pseudo-code, how you would implement solving this problem using the implicit method in time given an initial condition and boundary conditions of zero along the edge of the domain? You do not need to implement the method, but you need to outline a complete method to evolve the solution in time.

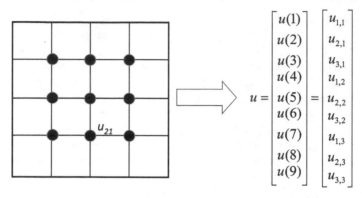

Fig. 4.2 The unknowns $u_{i,j}^{k+1}$ must be mapped to a vector of unknowns. Consider a fixed snapshot in time (so ignore the k superscripts for now). Starting a the bottom right corner, we move from left to right and assign each grid point to a location in the vector. Note this assignment (or ordering of your unknowns) will impact the structure of the matrix

   (d) How is the number of flops required different than the explicit case?

11. Consider solving the Bessel equation $x^2 y'' + xy' + (x^2 - 1) y = 0$ subject to the boundary conditions $y(1) = 1$, $y(15) = 0$ using numerical derivatives with $N = 280$ steps between $[1, 15]$. You should begin by rewriting the differential equation in the form

$$y'' + \frac{1}{x} y' + \left(1 - \frac{1}{x^2}\right) y = 0$$

   Next, use finite difference approximations to the derivatives and collect like terms to obtain a tridiagonal linear system for the unknown discrete values of y.

12. Solve $y'' + 2xy + 2y = 3e^{-x} - 2xe^{-x}$ subject to $y(-1) = e + 3/e$, $y(1) = 4/e$ using the second-order finite difference method as in the above problem. Consider $N = 10, 20, 40$ in $[-1, 1]$. Plot these solutions and the true solution $y = e^{-x} + 3e^{-x^2}$. Also, on a separate set of axes, plot the three error curves. Estimate the order of the truncation error for this method. Is it what you would expect?

4.3   LU Factorization and Applications

We now consider a way to reduce computational effort when having to solve multiple linear systems that have the same coefficient matrix, but different right hand side vectors. For example, consider solving $A\mathbf{x} = \mathbf{b}$ and then having to solve $A\mathbf{z} = \mathbf{c}$. Gauss elimination would yield the same computations the second time for the entries in A but those operations would still need to applied for the vector $\mathbf{c}$. This scenario is fairly common in practice. We demonstrate one example of how it arises in modeling below before giving the details of how to avoid doing the work on A more than once.

Specifically, this situation can arise when using an implicit discretization for the time derivative in a linear partial differential equation model (which are the underlying models in many industrial, large-scale simulation tools). A benefit is that stability is improved but the new formulation requires a linear solve at each time step. The backward Euler approximation is one such approach, which is implemented by simply considering all the values at the next time step. We demonstrate this on the heat equation.

Example 8 Consider the time dependent 1-D heat equation from Example 1 in Chap. 3 with an implicit time scheme and express the problem as a linear system.

Recall this model was given by,

$$\rho c u_t - K u_{xx} = f(x, t)$$

with the boundary conditions $u(0, t) = u(L, t) = 0\,°C$ and initial temperature $u_0(x) = u(x, 0)$. The idea is to evaluate all the terms in the discretized expressions for u_{xx} and f at $k + 1$.

$$\rho c \left[\frac{u_i^{k+1} - u_i^k}{\Delta t} \right] - K \left[\frac{u_{i-1}^{k+1} - 2u_i^{k+1} + u_{i+1}^{k+1}}{\Delta x^2} \right] = f_i^{k+1}. \qquad (4.12)$$

Multiplying through by $\Delta t / (\rho c)$ and rearranging, we arrive at a linear system of the form $A\mathbf{u}^{k+1} = \mathbf{b}$ of the form (for N = 5 interior nodes):

$$\begin{bmatrix} 1+2H & -H & & & \\ -H & 2H & -H & & \\ & -H & 2H & -H & \\ & & -H & 2H & -H \\ & & & -H & 2H \end{bmatrix} \begin{bmatrix} u_1^{k+1} \\ u_2^{k+1} \\ u_3^{k+1} \\ u_4^{k+1} \\ u_5^{k+1} \end{bmatrix} = \Delta x^2 \begin{bmatrix} u_1^k + \frac{\Delta t}{\rho c} t f_1 \\ u_2^k + \frac{\Delta t}{\rho c} t f_2 \\ u_3^k + \frac{\Delta t}{\rho c} t f_3 \\ u_4^k + \frac{\Delta t}{\rho c} t f_4 \\ u_5^k + \frac{\Delta t}{\rho c} t f_5 \end{bmatrix}$$

and $H = \left(\frac{K \Delta t}{\rho c \Delta x^2} \right)$. So, if $\mathbf{u}^0$ is our discrete initial condition vector we could define a right hand side vector, and solve $A\mathbf{u}^1 = \mathbf{b}$ to get the vector of discrete approximations $\mathbf{u}^1$ to $u(x, t^1)$ at the nodes. Then the process would be repeated with $\mathbf{u}^1$ to define a *new right hand side* vector $\mathbf{b}$ to solve for $\mathbf{u}^2$ but of course using the same matrix A. This would be repeated until we reached the final time. Performing Gauss elimination at each time step would be roughly $\mathcal{O}(N^3)$ work each time if we have N interior nodes.

It has already been mentioned that Gauss elimination could be used to solve several systems with the same coefficient matrix but different right-hand sides simultaneously. Essentially the only change is that the operations performed on the elements of the right-hand side *vector* must now be applied to the *rows* of the right-hand side *matrix*. Practically, it makes sense to avoid repeating all the work of the original solution in order to solve the second system. This can be achieved by keeping track of the multipliers used in the Gauss elimination process the first time. This leads us to what is called *LU factorization* of the matrix A in which we obtain two matrices L, a lower triangular matrix, and U, an upper triangular matrix, with the property that

$$A = LU \tag{4.13}$$

This process requires no more than careful storage of the multipliers used in Gauss elimination. To see this, consider the first step of the Gauss elimination phase where we generate the matrix

$$A^{(1)} = \begin{bmatrix} a_{11} & a_{12} & a_{13} & \cdots & a_{1n} \\ 0 & a'_{22} & a'_{23} & \cdots & a'_{2n} \\ 0 & a'_{32} & a'_{33} & \cdots & a'_{3n} \\ \vdots & \vdots & \vdots & \vdots & \vdots \\ 0 & a'_{n2} & a'_{n3} & \cdots & a'_{nn} \end{bmatrix}$$

in which

$$a'_{jk} = a_{jk} - \frac{a_{j1}}{a_{11}} a_{1k} = a_{jk} - m_{j1} a_{1k}$$

See (4.6) and (4.7). Next, use the multipliers to define

$$M_1 = \begin{bmatrix} 1 & 0 & \cdots & \cdots & 0 \\ m_{21} & 1 & 0 & \cdots & 0 \\ m_{31} & 0 & 1 & & \\ \vdots & \vdots & & \ddots & \\ m_{n1} & 0 & & & 1 \end{bmatrix} \tag{4.14}$$

Multiplying any matrix by M_1 results in adding m_{j1} times the first row to the jth row. Therefore, multiplying $A^{(1)}$ by M_1 would exactly reverse the elimination that has been performed, in other words

$$M_1 A^{(1)} = A \tag{4.15}$$

A similar equation holds for each step of the elimination phase. It follows that denoting the final upper triangular matrix produced by Algorithm 3 by U and defining the lower triangular matrix

$$
L = \begin{bmatrix}
1 & & & & & \\
m_{21} & 1 & & & & \\
m_{31} & m_{32} & 1 & & & \\
m_{41} & m_{42} & m_{43} & \ddots & & \\
\vdots & \vdots & \vdots & & 1 & \\
m_{n1} & m_{n2} & m_{n3} & \cdots & m_{n,n-1} & 1
\end{bmatrix}
\tag{4.16}
$$

where the m_{ji} are just the multipliers used in Algorithm 7, we get

$$
LU = A
$$

as desired.

As an algorithm this corresponds to simply overwriting the subdiagonal parts of A with the multipliers so that both the upper and lower triangles can be stored in the same local (matrix) variable. The entire process is shown in Algorithm 3.

Algorithm 7 *LU factorization of a square matrix*

> *Input* $n \times n$ matrix A
> *Factorization*
>   for i=1:n-1
>   for j=i+1:n
> $a_{ji} := \dfrac{a_{ji}}{a_{ii}}$;
>   for k=i+1:n
> $a_{jk} := a_{jk} - a_{ji} a_{ik}$
>   end
>   end
>   end
> *Output*   modified matrix A containing upper and lower triangular factors
> L is the lower triangle of A with unit diagonal elements
> U is the upper triangle of A

Note that this algorithm is really the same as Algorithm 3 except that the multipliers are stored in the lower triangle and the operations on the right-hand side of our linear system have been omitted. Using the LU factorization to solve $A\mathbf{x} = \mathbf{b}$ can be done in three steps:

1. Factor the matrix $A = LU$ as in Algorithm 7
2. Solve the lower triangular system $L\mathbf{z} = \mathbf{b}$ using *forward substitution*
3. Solve the upper triangular system $U\mathbf{x} = \mathbf{z}$ using back substitution as in Algorithm 4

Note that we have

$$Ax = LUx = Lz = b$$

so that we are obtaining the solution to our original problem. We can also see that the forward substitution algorithm is the natural equivalent of back substitution, but the divisions are not needed since the diagonal entries of L are all one. In terms of efficiency, although the total numbers of floating-point operations are the same as the original Gauss algorithm, with the factorization in hand, each subsequent solve only requires backward and forward elimination. So, it will take $\mathcal{O}(n^3)$ to get the LU and then only $\mathcal{O}(n^2)$ for each additional solve with new right-hand sides.

One of the downfalls with using the Gauss elimination approach is the accumulation of roundoff errors. Pivoting can improve matters, but even then, if the system is ill-conditioned, large relative errors can result. The LU factorization can help improve the solution.

To see how, we consider the vector of *residuals* for a computed solution $\tilde{x}$ to a system $Ax = b$. Define the residual vector by

$$\mathbf{r} = \mathbf{b} - A\tilde{\mathbf{x}} \tag{4.17}$$

The components of $\mathbf{r}$ given by

$$r_k = b_k - (a_{k1}\tilde{x}_1 + a_{k2}\tilde{x}_2 + \cdots + a_{kn}\tilde{x}_n)$$

for $k = 1, 2, \ldots, n$ provide a measure (although not always a very good one) of the extent to which we have failed to satisfy the equations. If we could also solve the system

$$Ay = r \tag{4.18}$$

then, by adding this solution $\mathbf{y}$ to the original computed solution $\tilde{\mathbf{x}}$, we obtain

$$A(\tilde{x} + y) = A\tilde{x} + Ay = A\tilde{x} + r$$
$$= A\tilde{x} + b - A\tilde{x} = b$$

In other words, if we can compute $\mathbf{r}$ and solve (4.18) *exactly*, then $\tilde{x} + \mathbf{y}$ is the exact solution of the original system. Of course, in practice, this is not possible but we would hope that $\tilde{x} + \tilde{y}$ will be closer to the true solution. (Here $\tilde{y}$ represents the computed solution of (4.18).) This suggests a possible iterative method, known as *iterative refinement,* for solving a system of linear equations.

Unfortunately, however, we do not know the residual vector at the time of the original solution. That is to say that (4.18) must be solved *after* the solution $\tilde{x}$ has been obtained. So what is the advantage? Consider the iterative refinement process that we discussed above. The initial solution $\tilde{x}$ is computed using $O\left(n^3\right)$ floating-point operations. Then the residual vector is computed as

$$\mathbf{r} = \mathbf{b} - A\tilde{\mathbf{x}}$$

after which the correction can be computed by solving

$$L\mathbf{z} = \mathbf{r}$$
$$U\mathbf{y} = \mathbf{z}$$

and setting

$$\mathbf{x} = \tilde{\mathbf{x}} + \mathbf{y}$$

The important difference is that the factorization does not need to be repeated if we make use of the *LU* factorization.

Example 9 Solve the system

$$\begin{bmatrix} 7 & -7 & 1 \\ -3 & 3 & 2 \\ 7 & 7 & -7 \end{bmatrix} \begin{bmatrix} x_1 \\ x_2 \\ x_3 \end{bmatrix} = \begin{bmatrix} 1 \\ 2 \\ 7 \end{bmatrix}$$

using LU factorization and iterative refinement.

First we note that this is the same system that we used in Example 4. From the multipliers used there, we obtain the factorization

$$\begin{bmatrix} 7 & -7 & 1 \\ -3 & 3 & 2 \\ 7 & 7 & -7 \end{bmatrix} = \begin{bmatrix} 1 & 0 & 0 \\ -0.4286 & 1 & 0 \\ 1 & -70000 & 1 \end{bmatrix} \begin{bmatrix} 7 & -7 & 1 \\ 0 & 14 & -8 \\ 0 & 0 & 169994 \end{bmatrix}$$

Solving $L\mathbf{z} = [1, 2, 7]'$ we obtain $\mathbf{z} = [1, 2.4286, 170008]'$ which is of course the right-hand side at the end of the elimination phase in Example 4. Now solving $U\mathbf{x} = \mathbf{z}$, to get the previously computed solution

$$\tilde{\mathbf{x}} = \begin{bmatrix} 1.2143 \\ 1.2143 \\ 1.0001 \end{bmatrix}.$$

The residuals are now given by

$$\mathbf{r} = \begin{bmatrix} 1 \\ 2 \\ 7 \end{bmatrix} - \begin{bmatrix} 7 & -7 & 1 \\ -3 & 3 & 2 \\ 7 & 7 & -7 \end{bmatrix} \begin{bmatrix} 1.2143 \\ 1.2143 \\ 1.0001 \end{bmatrix} = \begin{bmatrix} -0.0001 \\ -0.0002 \\ -2.9995 \end{bmatrix}$$

For the iterative refinement, we first solve

$$\begin{bmatrix} 1 & 0 & 0 \\ -0.4286 & 1 & 0 \\ 1 & -70000 & 1 \end{bmatrix} \mathbf{z} = \begin{bmatrix} -0.0001 \\ -0.0002 \\ -2.9995 \end{bmatrix}$$

to obtain $z = [-0.0001, -0.0002, -11.0004]'$ Then solving $U\mathbf{y} = \mathbf{z}$ yields $\mathbf{y} = [-0.2143, -0.2413, -0.0001]'$ and updating the previous solution gives

$$\mathbf{x} = \begin{bmatrix} 1.2143 \\ 1.2143 \\ 1.0001 \end{bmatrix} + \begin{bmatrix} -0.2143 \\ -0.2143 \\ -0.0001 \end{bmatrix} = \begin{bmatrix} 1 \\ 1 \\ 1 \end{bmatrix},$$

which is the exact solution in this case.

Although no pivoting was used in Example 9, partial pivoting can be included in the LU factorization algorithm. It is necessary to keep track of the order in which the rows were used as pivots in order to adjust the right-hand side appropriately. We shall not discuss the details of this implementation here.

When pivoting is used the factorization becomes equivalent to finding lower and upper triangular factors, and a *permutation matrix P* such that

$$PLU = A$$

LU factorization with partial pivoting is available in SciPy's linalg module as the function lu. It can return either two or three matrices. (With the argument permute_l = True, it returns two outputs; the upper triangular factor and a "lower triangular" factor with its rows interchanged. With argument permute_l = False) the function returns three arguments; the lower triangular factor is genuinely a lower triangle and the first argument is the permutation matrix.

For example with the same matrix as we used in Example 9, we can use

```
>>> L, U = lu (A, permute_l = True)
```

to obtain the output

```
>>> print (L)
[[   1.0000e+00    0.0000e+00    0.0000e+00]
 [  -4.2857e-01    1.1895e-17    1.0000e+00]
 [   1.0000e+00    1.0000e+00    0.0000e+00]]

>>> print (U)
[[   7.        -7.          1.      ]
 [   0.        14.         -8.      ]
 [   0.         0.          2.4286]]
```

where we see that the second and third rows of L have been interchanged to reflect the pivoting strategy. Note that

$$
\begin{bmatrix} 1.0000 & 0.0000 & 0.0000 \\ -0.42857 & 1.1895 \times 10^{-17} & 1.0000 \\ 1.0000 & 1.0000 & 0.0000 \end{bmatrix} \begin{bmatrix} 7.0000 & -7.0000 & 1.0000 \\ 0 & 14.0000 & -8.0000 \\ 0 & 0 & -2.4286 \end{bmatrix}
$$

$$
= \begin{bmatrix} 7.0000 & -7.0000 & 1.0000 \\ -3.0000 & 3.0000 & 2.0000 \\ 7.0000 & 7.0000 & -7.0000 \end{bmatrix}
$$

which is the original matrix A with small roundoff errors.

Alternatively, the command

```
>>> P, L, U = lu (A, permute_l = False)
```

yields

```
>>> print (P)
[[ 1.   0.   0.]
 [ 0.   0.   1.]
 [ 0.   1.   0.]]
```

```
>>> print (L)
[[   1.0000e+00    0.0000e+00    0.0000e+00]
 [   1.0000e+00    1.0000e+00    0.0000e+00]
 [  -4.2857e-01    1.1895e-17    1.0000e+00]]
```

```
>>> print (U)
[[   7.       -7.       1.     ]
 [   0.       14.      -8.     ]
 [   0.        0.       2.4286]]
```

The permutation matrix P carries the information on the order in which the rows were used. This time, we see that the product of L and U is

```
>>> print (L.dot (U))
[[ 7. -7.   1.]
 [ 7.  7.  -7.]
 [ 3.  3.   2.]]
```

which is the original matrix A with its second and third rows interchanged. This interchange is precisely the effect of multiplying by the permutation matrix P:

```
>>> print (P.dot (A))
[[ 7. -7.   1.]
 [ 7.  7.  -7.]
 [-3.  3.   2.]]
```

To complete the solution of the system in Example 9 we can then find z such that $Lz = Pb$ and x such that $Ux = z$. These operations yield the correct answers to full machine accuracy in this case.

Exercises

1. Obtain LU factors of

$$A = \begin{bmatrix} 7 & 8 & 8 \\ 6 & 5 & 4 \\ 1 & 2 & 3 \end{bmatrix}$$

using four decimal place arithmetic. Use these factors to solve the following system of equations

$$A \begin{bmatrix} x \\ y \\ z \end{bmatrix} = \begin{bmatrix} 7 \\ 5 \\ 2 \end{bmatrix}$$

with iterative refinement.

2. Write a program to perform forward substitution. Test it by solving the following system

$$\begin{bmatrix} 1 & & & & \\ 1 & 1 & & & \\ 2 & 3 & 1 & & \\ 4 & 5 & 6 & 1 & \\ 7 & 8 & 9 & 0 & 1 \end{bmatrix} \begin{bmatrix} x_1 \\ x_2 \\ x_3 \\ x_4 \\ x_5 \end{bmatrix} = \begin{bmatrix} 0.1 \\ 0.4 \\ 1.6 \\ 5.6 \\ 8.5 \end{bmatrix}$$

3. Write a program to solve a square system of equations using the built-in LU factorization function and your forward and back substitution programs. Use these to solve the Hilbert systems of dimensions 6 through 10 with solution vectors all ones.

4. For the solutions of Exercise 3, test whether iterative refinement offers any improvement.

5. Investigate what a Vandermonde matrix is (and we will see more about this in Chap. 6). Solve the 10×10 Vandermonde system $V\mathbf{x} = \mathbf{b}$ where $v_{ij} = i^{j-1}$ and $b_i = (i/4)^{10}$ using LU factorization.

6. Implement the model problem from Example 8, using the implicit discretization of the heat equation. Compare the computational effort of using Gauss elimination inside the time stepping loop compared to using the LU factorization calculated outside the time loop. Consider using a right hand side of all ones and experiment with different choices for Δt, Δx and finals times. As your final time gets larger, what would you expect the shape of your solution to look like?

4.4 Iterative Methods

The tridiagonal algorithm presented in Sect. 4.2 showed that the computational cost in solving $A\mathbf{x} = \mathbf{b}$ could be reduced if the underlying matrix had a known structure. If the dimension of the system is large and if the system is *sparse* (meaning A has many zero elements) then iterative methods can be attractive over direct methods. Two of these iterative methods are presented in this section. These two form the basis of a family of methods which are designed either to accelerate the convergence or to suit some particular computer architecture.

The methods can be viewed in terms of *splitting* A into a sum of its parts. Specifically, $A = L + U + D$ where L and U are triangular and are the strictly lower and upper parts of A (not to be confused with the LU decomposition). Here D a diagonal matrix with the diagonal of A on its diagonal. Specifically,

$$l_{ij} = \begin{cases} a_{ij} & \text{if } i > j \\ 0 & \text{otherwise} \end{cases}$$

$$d_{ii} = a_{ii}, \quad d_{ij} = 0 \text{ otherwise}$$

$$u_{ij} = \begin{cases} a_{ij} & \text{if } i < j \\ 0 & \text{otherwise} \end{cases}$$

The methods can also be viewed as a rearrangement of the original system of equations in the form

$$
\begin{aligned}
x_1 &= [b_1 - (a_{12}x_2 + a_{13}x_3 + \cdots + a_{1n}x_n)]/a_{11} \\
x_2 &= [b_2 - (a_{21}x_1 + a_{23}x_3 + \cdots + a_{2n}x_n)]/a_{22} \\
&\cdots \quad \cdots \quad \cdots \quad \cdots \quad \cdots \quad \cdots \\
x_n &= [b_n - (a_{n1}x_1 + a_{n2}x_2 + \cdots + a_{n,n-1}x_{n-1})]/a_{nn}.
\end{aligned}
\tag{4.19}
$$

This is actually the result of solving the ith equation for x_i in terms of the remaining unknowns. There is an implicit assumption here that all diagonal elements of A are nonzero. (It is always possible to reorder the equations of a nonsingular system to ensure this condition is satisfied.) This rearrangement of the original system lends itself to an iterative treatment.

For the *Jacobi iteration*, we generate the next estimated solution vector from the current one by substituting the current component values in the right-hand side of (4.19) to obtain the next iterates. In matrix terms, we set

$$\mathbf{x}^{(k+1)} = D^{-1}\left[\mathbf{b} - (L + U)\mathbf{x}^{(k)}\right] \tag{4.20}$$

where the superscript here represents an iteration counter so that $\mathbf{x}^{(k)}$ is the kth (vector) iterate. In component terms, we have

$$
\begin{aligned}
x_1^{(k+1)} &= \left[b_1 - \left(a_{12}x_2^{(k)} + a_{13}x_3^{(k)} + \cdots + a_{1n}x_n^{(k)}\right)\right]/a_{11} \\
x_2^{(k+1)} &= \left[b_2 - \left(a_{21}x_1^{(k)} + a_{23}x_3^{(k)} + \cdots + a_{2n}x_n^{(k)}\right)\right]/a_{22} \\
&\cdots \quad \cdots \quad \cdots \quad \cdots \quad \cdots \quad \cdots \\
x_n^{(k+1)} &= \left[b_n - \left(a_{n1}x_1^{(k)} + a_{n2}x_2^{(k)} + \cdots + a_{n,n-1}x_{n-1}^{(k)}\right)\right]/a_{nn}
\end{aligned}
\tag{4.21}
$$

Example 10 Perform the first three Jacobi iterations for the solution of the system

$$6x_1 + 3x_2 + 2x_3 = 26$$
$$2x_1 + 5x_2 + x_3 = 17$$
$$x_1 + x_2 + 4x_3 = 9$$

Rearranging the equations as in (4.19), we have

$$x_1 = \frac{26 - 3x_2 - 2x_3}{6}$$
$$x_2 = \frac{17 - 2x_1 - x_3}{5}$$
$$x_3 = \frac{9 - x_1 - x_2}{4}$$

and taking the initial guess $x^0 = 0$, we get the next iterates

$$x_1^1 = \frac{26}{6} = 4.3333$$
$$x_2^1 = \frac{17}{17} = 3.4000$$
$$x_3^1 = \frac{9}{4} = 2.2500$$

The next two iterations yield

$$x_1^2 = 1.8833, \quad x_2^2 = 1.2176, \quad x_3^2 = 0.3167$$

and

$$x_1^3 = 3.6194, \quad x_2^3 = 2.5833, \quad x_3^3 = 1.4750$$

which are (slowly) approaching the true solution $[3, 2, 1]^T$. In fact, eventually, $x^9 = [3.0498, 2.0420, 1.0353]^T$.

Clearly for this small linear system, an accurate solution with less effort could be found through a direct solve. The purpose of the example is simply to illustrate how the iteration works.

Stepping through the algorithm by hand may actually have brought a question to mind. Once $x_i^{(k+1)}$ has been obtained in the first of equations (4.21), why not use the most recent value in place of $x_i^{(k)}$ in the remainder of the updates in (4.21)? That's a great idea – the result is the *Gauss–Seidel iteration*:

$$x_1^{(k+1)} = \left[b_1 - \left(a_{12}x_2^{(k)} + a_{13}x_3^{(k)} + \cdots + a_{1n}x_n^{(k)} \right) \right] / a_{11}$$

$$x_2^{(k+1)} = \left[b_2 - \left(a_{21}x_1^{(k+1)} + a_{23}x_3^{(k)} + \cdots + a_{2n}x_n^{(k)} \right) \right] / a_{22}$$

$$\cdots \quad \cdots \quad \cdots \quad \cdots \quad \cdots \quad \cdots \tag{4.22}$$

$$x_n^{(k+1)} = \left[b_n - \left(a_{n1}x_1^{(k+1)} + a_{n2}x_2^{(k+1)} + \cdots + a_{n,n-1}x_{n-1}^{(k+1)} \right) \right] / a_{nn}$$

or, in matrix terms,

$$\mathbf{x}^{(k+1)} = D^{-1} \left[\mathbf{b} - \left(L\mathbf{x}^{(k+1)} + U\mathbf{x}^{(k)} \right) \right] \tag{4.23}$$

Note that (4.23) should be understood as an assignment of values from the right-hand side to the left where the entries of $\mathbf{x}^{(k+1)}$ are updated sequentially in their natural order.

Example 11 Repeat Example 10 using the Gauss–Seidel iteration.

With the same initial guess, the first iteration is now

$$x_1^1 = \frac{26}{6} = 4.3333$$

$$x_2^1 = \frac{17 - 2(4.3333)}{5} = 1.6667$$

$$x_3^1 = \frac{9 - 4.3333 - 1.6667}{4} = 0.7500$$

The next two iterations then produce the estimates

$$x_1^2 = 3.25, \quad x_2^2 = 1.95, \quad x_3^2 = 0.95$$
$$x_1^3 = 3.0417, \quad x_2^3 = 1.9933, \quad x_3^3 = 0.9912$$

which are much closer to the true solution than the Jacobi iterates for the same computational effort.

Either of the iterations (4.21) or (4.22) uses roughly n^2 multiplications and n divisions per iteration. Comparing this with the operation counts for Gauss elimination, we see that these iterative methods are likely to be computationally less expensive if they will converge in fewer than about $n/3$ iterations. For large sparse systems this may be the case.

With any iterative method, the question arises, under what conditions will the method converge? The examples above showed convergence, but unfortunately they are not guaranteed to behave that way. Consider a slight modification to Example 11 by changing the first coefficient from a 6 to a 1 in the first equation.

Now the first two Gauss–Seidel iterations give us

$$x_1^1 = 11.0000, \quad x_2^1 = -1.0000, \quad x_3^1 = -0.25$$
$$x_1^2 = 14.5000, \quad x_2^2 = -2.3500, \quad x_3^2 = -0.7875$$

which are clearly diverging from the solution.

There is a critical difference between these two systems. In the original ordering of the equations, the matrix of coefficients is *diagonally dominant* which is to say that each diagonal element is greater than the (absolute) sum of the other entries in its row, or

$$|a_{ii}| > \sum_{j \neq i} |a_{ij}|$$

for each i. For our example, we have $6 > 3 + 2, 5 > 2 + 1$, and $4 > 1 + 1$ whereas in the modified system the first row gives $1 < 3 + 2$.

The simplest conditions under which both Jacobi and Gauss–Seidel iterations can be proved to converge is when the coefficient matrix is diagonally dominant. The details of these proofs are beyond our present scope. They are essentially multivariable versions of the conditions for convergence of fixed-point (function) iteration which will be discussed for a single equation in the next chapter. The diagonal dominance of the matrix ensures that the various (multivariable) derivatives of the iteration functions (4.20) and (4.23) are all less than one.

In general, the Gauss–Seidel method will converge faster than the Jacobi method. On the other hand, the updates of a Jacobi iteration can all be performed simultaneously, whereas for Gauss–Seidel, $x_2^{(k+1)}$ cannot be computed until after $x_1^{(k+1)}$ is known. This means the Jacobi method has the potential to be easily parallelizable while Gauss–Seidel does not. There are other iterative schemes which can be employed to try to take some advantage of the ideas of the Gauss–Seidel iteration in a parallel computing environment.

This point is readily borne out by the efficient implementation of the Jacobi iteration in Python/NumPy where we can take advantage of the matrix operations.

Program   Python code for fixed number of Jacobi iterations

```
import numpy as np

def jacobit(A, b, Nits):
    """

    Performs Function for computing 'Nits'
    iterations of the
    Jacobi method for Ax=b where
    'A' must be square.

    """
```

```
D = np.diag(A)
n = A.shape[0]
A_D = A - np.diag(D)   # This is L + U
x = np.zeros(n)
s = np.zeros((n, Nits))
for k in range(Nits):
    x = (b - A_D.dot(x)) / D
    s[:, k] = x

return s
```

Note the ease of implementation that NumPy's matrix arithmetic allows for the Jacobi iteration. The same simplicity could not be achieved for Gauss–Seidel because of the need to compute each component in turn. The implicit inner loop would need to be "unrolled" into its component form.

Example 12 Perform the first six Jacobi iterations for a randomly generated diagonally dominant 10×10 system of linear equations.

One way to generate a suitable system is

```
>>> np.random.seed(42)
```

```
>>> A = 10 * np.eye(10)
>>> A += np.random.rand(10, 10)
>>> b = np.sum(A, axis = 1)
```

Since the random numbers are all in $[0, 1]$, the construction of A guarantees its diagonal dominance. The right-hand side is chosen to make the solution a vector of ones. Using the program listed above we get the table of results below.

```
>>> S = jacobit(A, b, 6)
```

Iteration	1	2	3	4	5	6
x_1	1.4653	0.8232	1.0716	0.9714	1.0114	0.9954
x_2	1.2719	0.8900	1.0447	0.9822	1.0071	0.9971
x_3	1.3606	0.8492	1.0599	0.9761	1.0096	0.9962
x_4	1.3785	0.8437	1.0615	0.9753	1.0099	0.9961
x_5	1.3691	0.8528	1.0592	0.9764	1.0095	0.9962
x_6	1.4424	0.8310	1.0677	0.9729	1.0108	0.9957
x_7	1.4470	0.8178	1.0725	0.9710	1.0116	0.9954
x_8	1.4441	0.8168	1.0727	0.9709	1.0116	0.9953
x_9	1.4001	0.8358	1.0652	0.9738	1.0105	0.9958
x_{10}	1.4348	0.8345	1.0670	0.9733	1.0107	0.9957

The iterates are gradually approaching the imposed solution.

The Jacobi iteration is particularly easy to program, but the Gauss–Seidel is typically more efficient. The results of using five Gauss–Seidel iterations for the system generated in Example 12 are tabulated below.

Iteration	1	2	3	4	5
x_1	1.4653	0.9487	1.0021	1.0003	1.
x_2	1.2692	0.9746	0.9996	1.0002	1.
x_3	1.2563	0.9914	0.9982	1.0001	1.
x_4	1.2699	0.9935	0.9977	1.0001	1.
x_5	1.1737	1.0066	0.9986	1.	1.
x_6	0.9459	1.0092	1.0000	0.9999	1.
x_7	1.1264	1.0130	0.9996	0.9999	1.
x_8	0.9662	1.0047	0.9999	1.	1.
x_9	0.9719	1.0074	0.9999	0.9999	1.
x_{10}	0.9256	1.0016	1.0004	1.	1.

It is apparent that the Gauss–Seidel iteration has given greater accuracy much more quickly.

Exercises

1. Perform the first three Jacobi iterations on the system

$$
\begin{bmatrix}
4 & 1 & & & \\
1 & 4 & 1 & & \\
& 1 & 4 & 1 & \\
& & 1 & 4 & 1 \\
& & & 1 & 4
\end{bmatrix}
\begin{bmatrix}
x_1 \\ x_2 \\ x_3 \\ x_4 \\ x_5
\end{bmatrix}
=
\begin{bmatrix}
4.1 \\ 2.4 \\ 4.2 \\ 2.4 \\ 4.1
\end{bmatrix}.
$$

2. Repeat Exercise 1 for the Gauss–Seidel iteration. Which appears better?
3. Write a program to perform a fixed number of Gauss–Seidel iterations for the solution of a system $Ax = b$. Test it on the system of Exercise 1.
4. Generate a random diagonally dominant matrix of dimension 20×20 and right-hand sides for true solutions (a) $x_i = 1$ for all i and (b) $x_i = (-1)^i$. Perform the first ten iterations of both Jacobi and Gauss–Seidel for each case. Compare your results.
5. Modify your code for the Gauss–Seidel iteration to continue until the maximum absolute difference between components of $x^{(k)}$ and $x^{(k+1)}$ is smaller than a given tolerance. Use your program to solve your example of Exercise 4 with accuracy 10^{-6}.
6. A psychological experiment which is used to test whether rats can learn consists of (repeatedly) putting a rat in a maze of interconnecting passages. At each intersection the rat chooses a direction and proceeds to the next intersection. Food is placed at one or more of the exits from the maze. If the rat learns then it should get food more frequently than would result from random decisions.

In the simplest case the "maze" consists of a rectangular grid of passages and the food is placed at exits on one edge of the grid. In order to decide whether the rat learns, it is first necessary to determine the probability that the rat gets food as a result of purely random decisions. Specifically, suppose the maze consists of a 6×5 grid of passages with exits at each edge. Food is placed at each exit on the left hand (western) edge of the grid. Let's consider finding the probabilities P_{ij} that the rat finds food from the initial position (i, j) where we are using "matrix coordinates". An example grid is in Fig. 4.3 and is augmented with the "probabilities of success" for starting points at each of the exits.

Now, for random decisions, assume that the probability of success (that is getting food) starting from position (i, j) will be simply the average of the probabilities at the four neighboring points. This can be written as a simple system of equations:

$$P_{ij} = \frac{1}{4} \left(P_{i-1,j} + P_{i+1,j} + P_{i,j-1} + P_{i,j+1} \right)$$

where $i = 1, 2, \ldots, 6$, $j = 1, 2, \ldots, 5$ and the probabilities at the edges are as shown. The array of values would therefore be augmented with $P_{0j} = P_{7j} = P_{i6} = 0$ and $P_{i0} = 1$. (The apparent conflict at the corners is irrelevant since there is no passage leading to the corner at $(0, 0)$ for example.)

For our case this could be rewritten as a conventional system of 30 linear equations in the unknowns P_{ij}. However this system would be very *sparse*. That is, a high proportion of the entries in the coefficient matrix would be 0 meaning this is a good candidate for an iterative solver. Use the *Jacobi* or *Gauss–Seidel* methods to calculate the probabilities.

7. The model used in the simple case above (Exercise 6) is probably too simple. It is used to test whether rats learn by statistical comparison of the rat's performance with that which would be expected as a result of purely random decisions. To make a more extensive test of rat's intelligence, the basic model needs refining.

- The food location could be changed for later experiments
- The size of the grid could also be changed
- Diagonal passages could be added.

Modify your code for these situations.

8. (See Exercises 6 and 7 above) The decisions that the rat makes are unlikely to be entirely *random*. It is probably more likely that a rat will continue in a given direction than that it will change direction – and, in particular, a $180°$ turn would be much less likely. Modify your model and code for the following hypotheses

- The rat never reverses its direction at one intersection (that is, no $180°$ turns)
- Incorporate variable probabilities for different turns with greater turning angles having lower probabilities. In particular for a rectangular maze try a probability of 1/2 for straight ahead, 1/4 for each $90°$ turn and 0 for a $180°$ turn.

Fig. 4.3 Example grid for
the maze with food locations
indicated with a 1

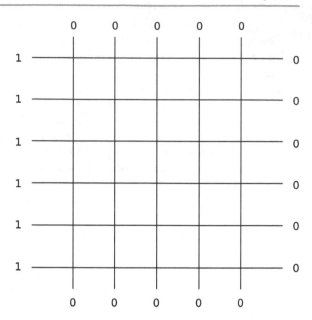

Try other values for the probabilities and extend the idea to diagonal passages,
too.

9. In this problem, we revisit the image distortion application in the exercises of
   Sects. 3.2 and 4.2. That exercise highlighted that explicit methods are often prone
   to stability problems. Implicit methods can help overcome this but require a linear
   solve at each time step. The backward Euler approximation is one such approach
   which simply considers all the values on the right hand side evaluated at the time
   $k + 1$ instead, as seen in Exercise 10 of Sect. 3.2. Often this system is too large to
   solve directly (i.e. with Gaussian Elimination) and since the matrix A is actually
   sparse (contains a lot of zeros) it is suitable for an iterative method, such as Jacobi
   or Gauss–Seidel.

   (a) What does the resulting linear system look like for the 2-D diffusion equation
       at each time step? How has the matrix structure changed in comparison to
       the 1-D case?
   (b) Determine (with pencil and paper) what the Jacobi iteration gives based on
       the matrix splitting for updating the i, jth component of the solution.
   (c) Since the matrix has mainly zeros, it would be inefficient to store the whole
       matrix and use Gaussian elimination (and the problem is large in size). The
       purpose of using the Jacobi method is that we do NOT need to store the
       matrix or even form the matrix splitting that the method is based on. You
       can manipulate the expression from Eq. (3.11) (with the terms on the right
       evaluated at $k + 1$) to represent the Jacobi iteration, which basically forms
       an iteration by diving by the diagonal entries, to solve for $u_{i,j}$ in terms of
       the others.

(d) Solve the 2-d diffusion equation for *image distortion* in Exercise 10 (Sect. 3.2) with the Jacobi iteration. Use your own image. Explore different time step sizes and draw some conclusions on the impact on the final solution.

Some remarks:

- This code should be written to take in an initial value $u(0)$ and solve the diffusion equation forward in time.
- Do NOT confuse the Jacobi iterates with u^{k+1} and u^k which represent solutions in TIME.
- To simplify the coding (and because we are applying this to an image, which is a non-physical problem) assume $h = 1$ in your coding.
- For points along the boundary, we avoid complexity by placing zeros all around the solution (i.e. assume zero boundary conditions).
- Because Jacobi requires an initial guess to get the iteration started, it is common to use the solution at the previous time step.

4.5 Linear Least Squares Approximation

We now return to one of the main motivating themes of this text; building mathematical models to understand the world around us and using scientific computing to find solutions. Models are often used to make predictions, sometimes created to try to match up well to data. The data may be experimental, historical, or may even be output from a complicated, computationally expensive function. In any case, one wants to trust that the model is a good predictor for other points of interest that are not yet measured or known. Unfortunately, since models are just that–models, they often are created based on a wide range of assumptions and simplifications that lead to unknown model *parameters*.

For example, suppose a team of safety analysts is trying to understand the distance of a car traveling at a constant speed when a driver is alerted that they need to stop (for example, when they see an animal about to run into the road). They come up with the following relationships based on a variety of assumptions (some based on physics and some based on observing and summarizing human behavior).

1. The reaction distance, that is the distance the car travels when a person realizes they need to stop and when they actually apply the brakes, is proportional to the speed.
2. The stopping distance, that is the distance a car travels once the brakes are applied, is proportional to the square of the speed.

These two assumptions together imply that if x is the speed of the car, then the total stopping distance from the time the person was alerted would be $D(x) = a_1 x + a_2 x^2$.

Next suppose they consider an experiment in which a car traveled at a fixed speed until the driver was told to stop and then the distance the car traveled from that moment until it completely stopped was recorded (See Table 4.1)

Their problem would then be to try to find the values of a_1 and a_2 so that their model fit as closely to the data as possible. Then, they may be able to predict the braking distance of a person driving at 48 km/h using the model. We'll go through the process of finding the unknown model parameter in an example later.

For now, let's consider the general problem; finding a polynomial of a certain degree, $p(x) = a_0 + a_1 x + a_2 x^2 + \ldots a_M x^M$ that agrees with a set of data,

$$D = \{(x_1, y_1), (x_2, y_2), \ldots, (x_N, y_N)\}.$$

Our problem can now be posed as to find the coefficients $a_0, a_1, \ldots a_M$ so that the polynomial agrees with the data as closely as possible at $x_1, x_2, \ldots, x_N$. Mathematically, this is usually posed as a least-squares approximation. That is, the problem is

$$\min_{\mathbf{a}} \sum_{i=1}^{N} (y_i - p(x_i))^2 \tag{4.24}$$

where $\mathbf{a}$ is a vector containing the coefficients in p. At first glance it may not be clear at all why this is called *linear* least squares but if the x_i are distinct then evaluation of $p(x)$ at those points actually leads to a linear system since

$$p(x_1) = a_0 + a_1 x_1 + a_2 x_1^2 + a_3 x_1^3 + \cdots + a_M x_1^M$$
$$p(x_2) = a_0 + a_1 x_2 + a_2 x_2^2 + a_3 x_2^3 + \cdots + a_M x_2^M$$
$$p(x_3) = a_0 + a_1 x_3 + a_2 x_3^2 + a_3 x_3^3 + \cdots + a_M x_3^M$$
$$\vdots$$
$$p(x_N) = a_0 + a_1 x_N + a_2 x_N^2 + a_3 x_N^3 + \cdots + a_M x_N^M$$

Table 4.1 Driving at the specified speed, this shows the distance the car traveled after the driver was told to stop

Speed (km/h)	Distance (m)
40	17
55	31
70	46
90	65
100	87

which can be written using matrix multiplication with **a** as

$$
\begin{bmatrix}
1 & x_1 & x_1^2 & x_1^3 & \cdots & x_1^M \\
1 & x_2 & x_2^2 & x_2^3 & \cdots & x_2^M \\
1 & x_3 & x_3^2 & x_3^3 & \cdots & x_3^M \\
 & & & \vdots & & \\
1 & x_N & x_N^2 & x_N^3 & \cdots & x_N^M
\end{bmatrix}
\begin{bmatrix}
a_0 \\ a_1 \\ a_2 \\ \vdots \\ a_M
\end{bmatrix}
$$

So our problem can actually be expressed as

$$
\min_{\mathbf{a}} \sum_{i=1}^{N} (y_i - p(x_i))^2 = (\mathbf{y} - A\mathbf{a})'(\mathbf{y} - A\mathbf{a}),
$$

where A is the coefficient matrix above and $\mathbf{y}$ is the vector of data measurements. We can expand this further using linear algebra. Recall that for a vector, say $\mathbf{x}$, that $\mathbf{x}'\mathbf{x} = \sum_i x_i^2$. So our problem expanded out is

$$
\min_{\mathbf{a}} F(\mathbf{a}) = \mathbf{a}'A'A\mathbf{a} - 2\mathbf{a}'A'\mathbf{y} + \mathbf{y}'\mathbf{y}
$$

When it comes to finding the minimum, we know from Calculus that this means setting the partial derivatives equal to zero. Here, those derivatives are $\frac{\partial F}{\partial a_i}$. This process will eventually lead to the system of equations given by

$$
A'A\mathbf{a} = A'\mathbf{y},
$$

known as the normal equations. The details of arriving at the normal equations for this discrete case is a bit beyond the scope of this book and left as an exercise (with some hints), but the theory is provided below for the continuous least squares problem.

There are multiple ways to solve the normal equations depending on the size of A. For example, if the x_i are distinct and if $N = M$, then A is a square, nonsingular matrix and therefore so is A' and so this problem reduces to solve $A\mathbf{a} = \mathbf{y}$. Usually, there are more data points than there are unknowns (that is $N > M$) which means that A is rectangular. Once again though, if the x_i are distinct then one can show that $A'A$ is a square, nonsingular matrix and the normal equations can be solved directly.

Example 13 Find the model parameters to fit the braking data in Table 4.1 in a least squares approximation.

Here the discrete model inputs (i.e. x_i) are taken as the five speeds and the model outputs (i.e. the y_i) are the stopping distances. Since the safety analysts' model has

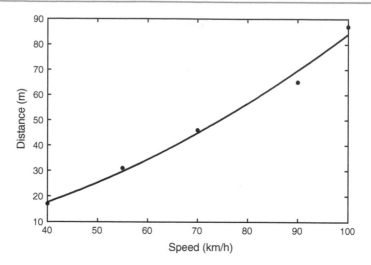

Fig. 4.4 Model for stopping distance plotted with experimental data

$a_0 = 0$, $D(x_i) = a_1 x_i + a_2^2 x_i^2$. The coefficient matrix has rows made up of x_i and x_i^2;

$$\begin{bmatrix} 40 & 1600 \\ 55 & 3025 \\ 70 & 4900 \\ 9 & 8100 \\ 100 & 10000 \end{bmatrix}.$$

Recall, A' is formed by transposing the rows and columns of A so that $A'A$ is the 2×2 matrix and the resulting linear system, $A'A\mathbf{a} = A'\mathbf{y}$, is given by

$$\begin{bmatrix} 27625 & 2302375 \\ 2302375 & 201330625 \end{bmatrix} \begin{bmatrix} a_1 \\ a_2 \end{bmatrix} = \begin{bmatrix} 20155 \\ 1742875 \end{bmatrix}$$

Solving this gives $a_1 = 0.1728$ and $a_2 = 0.0067$. So, the model can now be used to predict the stopping distance for intermediate speeds. For example to predict the stopping distance if driving 45 km/h we have $D(45) = 0.1728(45) + 0.0067(45^2) = 21.34$ m. Figure 4.4 shows the resulting model curve and data points plotted together. It is important to assess the quality of the model; how well does the model agree with the data? Calculating the absolute difference between the model at the speed values and the data (i.e. $|D(x_i) - y(i)|$) gives the following values 0.6007, 1.2873, 1.1689, 4.6655, 2.9130. So the biggest error is roughly 4.5 m that occurs when trying to match the stopping distance at 90 km/h. Whether or not the maximum error is acceptable, is of course up to the modeler (or perhaps the customer). They may decide to revisit their assumptions and come up with an entirely different model.

Similar problem formulation strategies apply when trying to approximate a function $f(x)$ with a polynomial. Recall from Sect. 2.4 that the continuous L_2 or *least squares*, metric is defined for the interval $[a, b]$ by

$$L_2(f, p) = \|f - p\|_2 = \sqrt{\int_a^b |f(x) - p(x)|^2 dx} \qquad (4.25)$$

The *continuous* least-squares approximation problem is to find the function p from the admissible set which minimizes this quantity. (In this case the "sum" of the squares of the errors is a continuous sum, or integral.)

To simplify notation, let $(L_2(f, p))^2$ be $F(a_0, a_1, \ldots, a_M) = F(\mathbf{a})$. Clearly the choice of $a_0, a_1, \ldots, a_M$ which minimizes F also minimizes $L_2(f, p)$. By definition

$$F(a_0, a_1, \ldots, a_M) = \int_a^b \left[f(x) - \left(a_0 + a_1 x + a_2 x^2 + \cdots + a_M x^M \right) \right]^2 dx. \qquad (4.26)$$

Setting each of the partial derivatives of F to zero, we obtain the following necessary conditions for the solution. These form a system of linear equations for the coefficients.

$$\frac{\partial F}{\partial a_0} = -2 \int_a^b \left[f(x) - \left(a_0 + a_1 x + a_2 x^2 + \cdots + a_M x^M \right) \right] dx = 0$$

$$\frac{\partial F}{\partial a_1} = -2 \int_a^b x \left[f(x) - \left(a_0 + a_1 x + a_2 x^2 + \cdots + a_M x^M \right) \right] dx = 0$$

$$\vdots \qquad \vdots \qquad \vdots \qquad \vdots \qquad \vdots \qquad \vdots$$

$$\frac{\partial F}{\partial a_M} = -2 \int_a^b x^M \left[f(x) - \left(a_0 + a_1 x + a_2 x^2 + \cdots + a_M x^M \right) \right] dx = 0$$

Dividing each of these by 2 and we can rewrite this system as

$$\begin{aligned}
c_0 a_0 + c_1 a_1 + \cdots + c_M a_M &= b_0 \\
c_1 a_0 + c_2 a_1 + \cdots + c_{M+1} a_M &= b_1 \\
c_2 a_0 + c_3 a_1 + \cdots + c_{M+2} a_M &= b_2 \\
\vdots \qquad \vdots \qquad \vdots \quad &= \vdots \\
c_M a_0 + c_{M+1} a_1 + \cdots + c_{2M} a_M &= b_M
\end{aligned} \qquad (4.27)$$

where the coefficients c_k and b_k for appropriate values of k are given by

$$c_k = \int_a^b x^k dx, \quad b_k = \int_a^b x^k f(x)\, dx \qquad (4.28)$$

In matrix terms the system is

$$
\begin{bmatrix}
c_0 & c_1 & c_2 & \cdots & c_M \\
c_1 & c_2 & \cdots & c_M & c_{M+1} \\
c_2 & \cdots & c_M & c_{M+1} & c_{M+2} \\
\cdots & & \cdots & & \cdots \\
c_M & c_{M+1} & c_{M+2} & \cdots & c_{2M}
\end{bmatrix}
\begin{bmatrix}
a_0 \\ a_1 \\ a_2 \\ \vdots \\ a_M
\end{bmatrix}
=
\begin{bmatrix}
b_0 \\ b_1 \\ b_2 \\ \vdots \\ b_M
\end{bmatrix}
\tag{4.29}
$$

Note the special structure of the matrix with constant entries on each "reverse diagonal". Such matrices are often called *Hankel matrices*.

In the discrete case that we investigated above, a similar system was obtained, the only difference being that the coefficients are defined by the discrete analogues of (4.28):

$$
c_k = \sum_{i=0}^{N} x_i^k, \quad b_k = \sum_{i=0}^{N} x_i^k f(x_i)
\tag{4.30}
$$

These are exactly the entries in $A'A$ and $A'\mathbf{y}$ if you consider $y_i = f(x_i)$. In either case therefore the least-squares approximation problem is reduced to a linear system of equations, the *normal equations*. The matrix of coefficients is necessarily nonsingular (provided the data points are distinct in the discrete case, with $M = N$) and so the problem has a unique solution. The reason this process is called *linear least-squares* is that the approximating function is a linear combination of the basis functions, in this case the monomials $1, x, x^2, \ldots, x^M$.

Example 14 Find the continuous least squares cubic approximation to $\sin x$ on $[0, \pi]$.

We must minimize

$$
F(a_0, a_1, a_2, a_3) = \int_0^\pi \left(\sin x - a_0 - a_1 x - a_2 x^2 - a_3 x^3 \right) dx
$$

The coefficients of the normal equations are given by

$$
c_k = \int_0^\pi x^k dx = \frac{\pi^{k+1}}{k+1}
$$

$$
b_k = \int_0^\pi x^k \sin x \, dx
$$

so that $b_0 = 2$, $b_1 = \pi$, $b_2 = \pi^2 - 4$, and $b_3 = \pi(\pi^2 - 6)$. The normal equations are therefore

$$
\begin{bmatrix}
\pi & \pi^2/2 & \pi^3/3 & \pi^4/4 \\
\pi^2/2 & \pi^3/3 & \pi^4/4 & \pi^5/5 \\
\pi^3/3 & \pi^4/4 & \pi^5/5 & \pi^6/6 \\
\pi^4/4 & \pi^5/5 & \pi^6/6 & \pi^7/7
\end{bmatrix}
\begin{bmatrix}
a_0 \\ a_1 \\ a_2 \\ a_3
\end{bmatrix}
=
\begin{bmatrix}
2 \\ \pi \\ \pi^2 - 4 \\ \pi(\pi^2 - 6)
\end{bmatrix}
$$

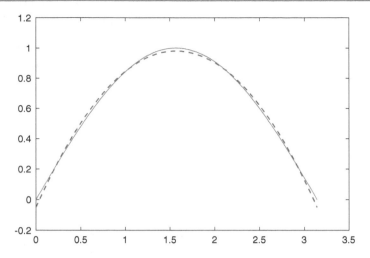

Fig. 4.5 $f(x) = sin(x)$ and the least squares cubic polynomial approximation (dashed line)

Applying any of our linear equation solvers (LU factorization or Gauss elimination) we obtain the solution

$a_0 = -0.0505, \; a_1 = 1.3122, \; a_2 = -0.4177, \; a_3 = 0$

so that the required least squares approximation is

$$\sin x \approx -0.0505 + 1.3122x - 0.4177x^2$$

which, for example, yields the approximations

$$\sin \pi/2 \approx 0.9802, \; \text{and} \; \sin \pi/4 \approx 0.7225$$

The plot of the approximation (dashed line) along with the true solution is shown in Fig. 4.5 and the error is show in Fig. 4.6

The Python (NumPy) function **polyfit** performs essentially these operations to compute discrete least-squares polynomial fitting. See Sect. 4.8.2 for details.

Unfortunately, the system of resulting linear equations for the method presented above tends to be ill-conditioned, particularly for high-degree approximating polynomials. For example, consider the continuous least-squares polynomial approximation over [0, 1] . The coefficients of the linear system will then be

$$h_{ij} = c\,(i + j - 1) = \int_0^1 x^{i+j-2} dx = \frac{1}{i + j - 1}$$

which is to say the coefficient matrix will be the Hilbert matrix of appropriate size. As mentioned earlier in the text, as the dimension increases, solutions of such Hilbert systems become increasingly incorrect.

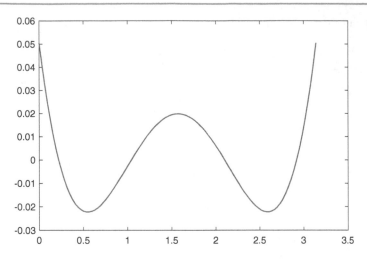

Fig. 4.6 Error in the least squares cubic approximation to $sin(x)$

The same ill-conditioning holds true even for discrete least-squares coefficient matrices so that alternative methods are needed to obtain highly accurate solutions. These alternatives use a different basis for the representation of our approximating polynomial. Specifically a basis is chosen for which the linear system will be well-conditioned and possibly sparse. The ideal level of sparsity would be for the matrix to be diagonal. This can be achieved this by using *orthogonal polynomials* as our basis functions.

Let $\phi_0, \phi_1, \ldots, \phi_M$ be polynomials of degrees $0, 1, \ldots, M$ respectively. These polynomials form a basis for the space of polynomials of degree $\leq M$. Our approximating polynomial p can then be written as a linear combination of these basis polynomials;

$$p(x) = a_0\phi_0(x) + a_1\phi_1(x) + \cdots + a_M\phi_M(x). \tag{4.31}$$

The problem can then be posed to minimize

$$F(a_0, a_1, \ldots, a_M) = \int_a^b [f(x) - (a_0\phi_0(x) + a_1\phi_1(x) + \cdots + a_M\phi_M(x))]^2\, dx$$

The same procedure applies–we set all partial derivatives of F to zero to obtain a system of linear equations for the unknown coefficients;

$$\begin{bmatrix} c_{00} & c_{01} & c_{02} & \cdots & c_{0M} \\ c_{10} & c_{11} & c_{12} & \cdots & c_{1M} \\ c_{20} & c_{21} & c_{22} & \cdots & c_{2M} \\ \cdots & \cdots & & \cdots & \\ c_{M0} & c_{M1} & c_{M2} & \cdots & c_{MM} \end{bmatrix} \begin{bmatrix} a_0 \\ a_1 \\ a_2 \\ \vdots \\ a_M \end{bmatrix} = \begin{bmatrix} b_0 \\ b_1 \\ b_2 \\ \vdots \\ b_M \end{bmatrix} \tag{4.32}$$

where

$$c_{ij} = c_{ji} = \int_a^b \phi_i(x)\phi_j(x)\,dx$$

$$b_i = \int_a^b f(x)\phi_i(x)\,dx$$

The key is to find a set of polynomials $\phi_0, \phi_1, \ldots$ such that each ϕ_i is of degree i and such that

$$c_{ij} = \int_a^b \phi_i(x)\phi_j(x)\,dx = 0 \qquad (4.33)$$

whenever $i \neq j$. In this case, the system (4.32) reduces to a diagonal system

$$
\begin{bmatrix}
c_{00} & 0 & 0 & \cdots & 0 \\
0 & c_{11} & 0 & \cdots & 0 \\
0 & 0 & c_{22} & \cdots & 0 \\
\cdots & & & \cdots & \\
0 & 0 & \cdots & 0 & c_{MM}
\end{bmatrix}
\begin{bmatrix}
a_0 \\ a_1 \\ a_2 \\ \vdots \\ a_M
\end{bmatrix}
=
\begin{bmatrix}
b_0 \\ b_1 \\ b_2 \\ \vdots \\ b_M
\end{bmatrix}
\qquad (4.34)
$$

which can easily be seen to give

$$a_j = \frac{b_j}{c_{jj}} = \frac{\int_a^b f(x)\phi_i(x)\,dx}{\int_a^b [\phi_i(x)]^2\,dx} \qquad (4.35)$$

Such polynomials are known as *orthogonal polynomials* over $[a, b]$. The members of such a set of orthogonal polynomials will depend on the specific interval $[a, b]$. Note that there are many classes of orthogonal polynomials and actually, other orthogonal functions could be used for approximation depending on the context.

We focus on one particular set of orthogonal polynomials called the *Legendre polynomials* $P_n(x)$. Multiplying orthogonal functions by scalar constants does not affect their orthogonality and so some normalization is needed. For the Legendre polynomials one common normalization is to set $P_n(1) = 1$ for each n. With this normalization the first few polynomials on the interval $[-1, 1]$ are

$$
\begin{aligned}
&P_0(x) = 1, \quad P_1(x) = x \\
&P_2(x) = \left(3x^2 - 1\right)/2 \\
&P_3(x) = \left(5x^3 - 3x\right)/2 \\
&P_4(x) = \left(35x^4 - 30x^2 + 3\right)/8
\end{aligned}
\qquad (4.36)
$$

Example 15 Find the first three orthogonal polynomials on the interval $[0, \pi]$ and use them to find the least-squares quadratic approximation to $\sin x$ on this interval.

The first one, ϕ_0 has degree zero and leading coefficient 1. As we define the subsequent polynomials, we normalize them so the leading coefficient is one. It follows that

$$\phi_0(x) = 1$$

Next $\phi_1(x)$ must have the form $\phi_1(x) = x + c$ for some constant c, and must be orthogonal to ϕ_0. That is

$$\int_0^\pi \phi_1(x)\,\phi_0(x)\,dx = \int_0^\pi (x+c)(1)\,dx = 0$$

It follows that $\pi^2/2 + c\pi = 0$ so that $c = -\pi/2$ and

$$\phi_1(x) = x - \frac{\pi}{2}$$

Next, let $\phi_2(x) = x^2 + ax + b$ and must choose a, b so that it is orthogonal to both ϕ_0 and ϕ_1. It is equivalent to force ϕ_2 to be orthogonal to both $1, x$. These conditions are

$$\int_0^\pi \left(x^2 + ax + b\right) dx = \frac{\pi^3}{3} + \frac{b\pi^2}{2} + c\pi = 0$$
$$\int_0^\pi x\left(x^2 + ax + b\right) dx = \frac{\pi^4}{4} + \frac{b\pi^3}{3} + \frac{c\pi^2}{2} = 0$$

giving $a = -\pi$ and $b = \pi^2/6$ so that

$$\phi_2(x) = x^2 - \pi x + \pi^2/6$$

The coefficients of the system (4.34) are then

$$c_{00} = \pi, \quad c_{11} = \frac{\pi^3}{12}, \quad c_{22} = \frac{\pi^5}{180}$$

$$b_0 = 2, \quad b_1 = 0, \quad b_2 = \frac{\pi^2}{3} - 4$$

giving the coefficients of the approximation

$$a_0 = \frac{2}{\pi}, \quad a_1 = \frac{0}{\pi^3/12}, \quad a_2 = \frac{\left(\pi^2 - 12\right)/3}{\pi^5/180}$$

Putting it all together, the approximation is

$$\sin x \approx \frac{2}{\pi} + \frac{60\pi^2 - 720}{\pi^5}\left(x^2 - \pi x + \pi^2/6\right)$$

which can be rewritten as

$$\sin x \approx -0.0505 + 1.3122x - 0.4177x^2.$$

The rearrangement simply allows us to see that this is exactly the same approximation derived in Example 14.

One immediate benefit of using an orthogonal basis for least-squares approximation is that additional terms can be added with little effort because the resulting system is diagonal. The previously computed coefficients remain valid for the higher-degree approximation and only the integration and division are needed to get the next coefficient. Moreover, the resulting series expansion that would result from continuing the process usually converges faster than a simple power series expansion, giving greater accuracy for less effort.

Keep in mind, there other functions that may be appropriate for least squares approximations instead of polynomials. For periodic functions, such as waves, including square or triangular waves, periodic basis functions are a suitable choice. On the interval $[-\pi, \pi]$, the functions $\sin(mx)$ and $\cos(nx)$ form an orthogonal family. The least-squares approximation using these functions is the *Fourier series*. The coefficients of this series represent the Fourier transform of the original function. We do not go into those types of problems here but hope the reader will investigate the wide range of possibilities when it comes to least squares problems since they arise frequently in practice.

Exercises

1. Duplicate the results in Example 14.
2. Find the least-squares cubic approximation to e^x on $[-1, 1]$.
3. Verify that $P_0, P_1, \ldots, P_4$ given by (4.36) are orthogonal over $[-1, 1]$.
4. Evaluate $\int_{-1}^{1} x^k e^x \, dx$ for $k = 0, 1, \ldots, 4$. Use these to obtain the least-squares quartic (degree 4) approximation to e^x on $[-1, 1]$. Plot this function, and your answer to Exercise 1, and e^x on the same set of axes. Also compare the error curves.
5. Compute the discrete least-squares quadratic approximation to e^x on $[-1, 1]$ using 21 equally spaced data points.
6. Repeat Exercise 4 for the quartic approximation. Compare the error curves.
7. Find the least-squares quartic approximations to $|x|$ on $[-1, 1]$ using

   (a) the discrete metric with data points at $-1 : 1/5 : 1$, and
   (b) the continuous least-squares metric.

   Plot the two approximations on the same set of axes.
8. Find the orthogonal polynomials of degree up to 4, each with leading coefficient 1, on $[0, 1]$.

9. Use the results of Exercise 7 to obtain the fourth degree least-squares approximation to arctan x on $[0, 1]$. Plot the error curve for this approximation.

10. The annual profits for a given company are shown below. Use a least squares approach to obtain a polynomial to predict the company's profits. Should the company close down next year? In two years? When should the company decide it is losing too much money?

year	profits (in thousands)
1	18
2	12
3	9
4	6

11. A biologist is modeling the trajectory of a frog using a motion sensor to collect the position of the frog as it leaps. She has the following positions;

height (cm)	distance (cm)
0	0
1.2	6.1
2.4	9.0
3.6	8.7
4.8	5.8
6	0

Find the quadratic polynomial that best fits the data. Use your quadratic model to predict the maximum height of the frog and the horizontal distance it might occur. Plot the data and the polynomial together.

12. Newton's Law of Cooling (or warming) states that the difference between an object's current temperature and the ambient temperature decays at a rate proportional to that difference. This results in an exponential decay model for the temperature difference as a function of time that could be modeled as

$$T(t) = T_E + (T_0 - T_E)e^{-kt}.$$

Here, $T(t)$ is the temperature of the object at time t, T_0 is the initial temperature, and T_E is the temperature of the environment (i.e. the ambient or room temperature).

Note that the model is not linear in its coefficients and so we need to modify the data and model (we address the nonlinearities and give more background on the model in the next chapter). First rearrange the equation as

$$T - T_E = (T_0 - T_E)e^{-kt}$$

and note that $T - T_E$ and $T_0 - T_E$ are both negative. If we multiply both sides by -1, we can use logarithms to obtain a linear least squares problem to obtain estimates of $T_E - T_0$ and k.

(a) Take a cold drink out of the refrigerator in a can or bottle–but don't open it yet. Gather data on the temperature of the drink container after 5 min, and at three minute intervals for another fifteen minutes. You will have six readings in all.

(b) Use linear least squares to estimate the temperature of your refrigerator and the constant k for this particular container. It is reasonable to suppose that the refrigerator temperature is the initial temperature T_0 of your container assuming it has been inside for some time.

(c) How long will it take your drink to reach $15.5\,°C$ or $60\,°F$? Check your answer by measuring the temperature after this time.

(d) What does your solution say the temperature of your refrigerator is? Check it's accuracy by testing the actual refrigerator.

(e) How could you modify this procedure if you do not know the ambient temperature?

(f) What would happen if that temperature was itself changing?

4.6   Eigenvalues

If you have had a course in applied linear algebra or differential equations, the notion of eigenvalues may be familiar. What may be less clear is why they are so important. Probably everyone reading this in fact uses an enormous eigenvalue problem solver many time a day. Google's page rank algorithm is, at its heart, a very large eigenvalue solver. At full size, the matrix system would have dimension in the many billions–but of course this matrix would be very sparse and most entries are irrelevant to any particular search.

A widely used tool in statistics and data science is Principal Component Analysis in which we identify the factors of most critical importance to some phenomenon of interest, such as identifying which aspects of a student's pre-college preparation are most important predictors of their likely degree completion, or which courses they should be placed in at the start of their college career. This is an eigenvalue problem. Closely related to this is identifying primary modes, for example the fundamental frequency in a vibrating structure–again an eigenvalue problem. Identifying such modes can be critically important in order to avoid resonant vibration in a bridge that could lead to its failure.

Biologists and environmental scientists are often concerned with population dynamics under different scenarios. The long term behavior of such population models typically depends on the eigenvalues of the underlying matrix and these allow determination of the likelihood of the population gradually dying out, or growing essentially unbounded or reaching a stable equilibrium within the carrying capacity of the local ecology.

In this section, we review some fundamental ideas about eigenvalues and focus on some iterative methods for computing them. Recall that λ is an *eigenvalue* of a matrix A if there exists a nonzero vector, an associated *eigenvector*, $\mathbf{v}$ such that

$$A\mathbf{v} = \lambda\mathbf{v} \tag{4.37}$$

In what follows, $(\lambda, \mathbf{v})$ is referred to as an eigenpair. Note that definition (4.37) implies that if λ is an eigenvalue of A then the system of equations

$$(A - \lambda I)\mathbf{v} = \mathbf{0}$$

has nontrivial (i.e. nonzero) solutions so that

$$\det(A - \lambda I) = 0. \tag{4.38}$$

Equation (4.38) represents a polynomial equation in λ which, in principle could be used to obtain the eigenvalues of A. To see this, consider a simple 2×2 matrix;

$$det\left(\begin{bmatrix} a & b \\ c & d \end{bmatrix} - \lambda I\right) = (a - \lambda)(d - \lambda) - bc = 0$$

This equation is called the *characteristic equation* of A.

Note that the roots of the characteristic equation are the eigenvalues of A, but solving for those roots directly is almost *never* a practical way to find eigenvalues when the size of A is anything more 2! Also observe that the eigenvalues of real matrices can be real or complex, and can be repeated. The *geometric multiplicity* of an eigenvalue is the dimension of the vector space of its associated eigenvectors. For an $n \times n$ matrix, the sum of the algebraic multiplicities of its eigenvalues is n. If the sum of the geometric multiplicities is less than n, the matrix is called defective.

Some basic facts which are useful in determining eigenvalues are

1. If A is real and symmetric, then its eigenvalues are real.
2. The trace of A, $\sum a_{ii}$ equals the sum of its eigenvalues.
3. The determinant of A, $\det A$ equals the product of its eigenvalues.

Eigenvalues can be used to measure how *ill-conditioned* a matrix is, something we discussed briefly early in the context of linear systems. A matrix is called ill-conditioned if its eigenvalues are spread far apart, measured in terms of a *condition number*. The most commonly used condition number for a matrix is defined by

$$\kappa(A) = \frac{|\lambda_{\max}|}{|\lambda_{\min}|} \tag{4.39}$$

where $\lambda_{\max}$ and $\lambda_{\min}$ are the (absolute) largest and smallest eigenvalues of A. Strictly speaking a matrix has many condition numbers, and "large" is not itself well-defined in this context. To get an idea of what "large" means, the condition number of the 6×6 Hilbert matrix is around 15×10^6 while that for the well-conditioned matrix

$$
\begin{bmatrix}
1 & 2 & 3 & 4 & 5 & 6 \\
2 & 2 & 3 & 4 & 5 & 6 \\
3 & 3 & 3 & 4 & 5 & 6 \\
4 & 4 & 4 & 4 & 5 & 6 \\
5 & 5 & 5 & 5 & 5 & 6 \\
6 & 6 & 6 & 6 & 6 & 6
\end{bmatrix}
$$

is approximately 100.

We focus on the *power method* for finding eigenvalues and associated eigenvectors and show how it can be used to estimate condition numbers. In its basic form the power method is a technique which will provide the *largest* (in absolute value) eigenvalue of a square matrix A.

The algorithm is based on the fact that if $(\lambda, \mathbf{v})$ form an eigen pair, then

$$
\frac{\mathbf{v}'A\mathbf{v}}{\mathbf{v}'\mathbf{v}} = \frac{\mathbf{v}'(\lambda\mathbf{v})}{\mathbf{v}'\mathbf{v}} = \lambda\frac{||\mathbf{v}||^2}{||\mathbf{v}||^2} = \lambda \tag{4.40}
$$

The ratio $\dfrac{\mathbf{v}'A\mathbf{v}}{\mathbf{v}'\mathbf{v}}$ is known as the *Rayleigh quotient*. We shall use the standard Euclidean norm $||\mathbf{v}||^2 = \mathbf{v} \cdot \mathbf{v} = \sum v_i^2$ where v_i is the ith component of $\mathbf{v}$.

Algorithm 8 *Basic power method using Rayleigh quotient*

Input Square matrix A, and initial nonzero vector $\mathbf{x}_0$; tolerance ε
Initialize $k = 0$
Repeat $\mathbf{v} := A\mathbf{x}_k$
$$\lambda_{k+1} := \frac{\mathbf{v}'\mathbf{x}_k}{||\mathbf{x}_k||^2}$$
$$\mathbf{x}_{k+1} := \mathbf{v}/||\mathbf{v}||$$
$$k = k + 1$$
until $|\lambda_k - \lambda_{k-1}| < \varepsilon$
Output Eigen pair $(\lambda_k, \mathbf{x}_k)$

The idea is that each time the vector is multiplied by the matrix A, the scale factor between $\mathbf{x}_k$ and $\mathbf{v}$ will get closer to the dominant eigenvalue $\lambda_{\max}$. If there is a dominant eigenvalue, then the convergence proof is straightforward given a good choice of $\mathbf{x}_0$. The details of the proof are omitted, but the basic idea is that the vector $\mathbf{x}_0$ can be written as a linear combination of eigenvectors corresponding to the different eigenvalues. As this vector is multiplied by increasing powers of the matrix

A, eventually the contribution corresponding to the largest eigenvalue will dominate. The result is that the Rayleigh quotient will converge to that largest eigenvalue.

In Algorithm 8, the renormalization step $\mathbf{x}_{k+1} := \mathbf{v}/\|\mathbf{v}\|$ is only needed to counteract the potential for enormous growth in the elements of the vectors generated. Simply setting $\mathbf{x}_{k+1} := A\mathbf{x}_k$ would lead to the same eigenvalue provided the vector components do not overflow prior to convergence. Note that,

$$\mathbf{x}_k = A^k \mathbf{x}_0 \tag{4.41}$$

which explains the origin of the name "power method" for this algorithm.

Variants of the algorithm use different ratios in place of the Rayleigh quotient. Other possibilities are the (absolute) largest components of the vector, the (absolute) sum of the vector components, the first elements of the vectors and so on. For vectors of moderate dimension, the Euclidean norm is easily enough computed that there is no great advantage in using these cheaper alternatives.

We note that the existence of a dominant eigenvalue for a real matrix necessarily implies that this dominant eigenvalue is *real*. The power method and its variants are therefore designed for finding real eigenvalues of real matrices or complex eigenvalues of complex matrices. They are not suited to the (common) situation of real matrices with complex eigenvalues.

Example 16 Find the largest eigenvalue of the matrix

$$A = \begin{bmatrix} 1 & 2 & 3 & 4 & 5 & 6 \\ 2 & 2 & 3 & 4 & 5 & 6 \\ 3 & 3 & 3 & 4 & 5 & 6 \\ 4 & 4 & 4 & 4 & 5 & 6 \\ 5 & 5 & 5 & 5 & 5 & 6 \\ 6 & 6 & 6 & 6 & 6 & 6 \end{bmatrix}$$

using the power method.

With the arbitrary initial guess $\mathbf{x}_0 = [1, 2, 3, 4, 5, 6]^{\mathrm{T}}$, the results of the first few iterations are

λ	Approximate associated eigenvector
25	$[0.3592, 0.3632, 0.3750, 0.3987, 0.4382, 0.4974]^{\mathrm{T}}$
27.6453	$[0.3243, 0.3373, 0.3634, 0.4030, 0.4571, 0.5269]^{\mathrm{T}}$
27.7209	$[0.3300, 0.3417, 0.3656, 0.4025, 0.4540, 0.5220]^{\mathrm{T}}$
27.7230	$[0.3291, 0.3410, 0.3652, 0.4026, 0.4545, 0.5229]^{\mathrm{T}}$
27.7230	$[0.3292, 0.3411, 0.3653, 0.4026, 0.4545, 0.5227]^{\mathrm{T}}$

We see that the eigenvalue estimates have already settled to three decimals and the components of the associated eigenvector are also converging quickly.

The power method seems to be reasonable for finding the dominant eigenvalue of a matrix. The good news is that it can be modified to find other eigenvalues. The next easiest eigenvalue to obtain is the (absolute) smallest one. To see why, note that for a nonsingular matrix A, the eigenvalues of A^{-1} are the reciprocals of those of A since if $(\lambda, \mathbf{v})$ are an eigen pair of A, then

$$A\mathbf{v} = \lambda\mathbf{v}.$$

Premultiplying by A^{-1} gives $\mathbf{v} = \lambda A^{-1}\mathbf{v}$, or

$$\frac{1}{\lambda}\mathbf{v} = A^{-1}\mathbf{v}.$$

It follows that the smallest eigenvalue of A is the reciprocal of the largest eigenvalue of A^{-1}. This largest eigenvalue of A^{-1} could be computed using the power method – except we do not have A^{-1}. The following technique, the *inverse iteration*, can be implemented by taking advantage of the LU factorization of A to avoid a repeated linear solve with different right hand sides.

Note, at each iteration, finding $\mathbf{v} = A^{-1}\mathbf{x}_k$, is equivalent to solving the system

$$A\mathbf{v} = \mathbf{x}_k$$

which can be done with forward and back substitution if the LU factors of A are known. The algorithm for inverse iteration can be expressed as;

Algorithm 9 *Inverse iteration using LU factorization*

> *Input*   Matrix A, initial nonzero vector $\mathbf{x}_0$; tolerance ε
> *Initialize*   Find LU factorization $LU = A$
>   Set $k = 0$
> *Repeat*   Solve $L\mathbf{w} = \mathbf{x}_k$
>   Solve $U\mathbf{v} = \mathbf{w}$
> $\lambda_k := \dfrac{\mathbf{v}'\mathbf{x}_k}{||\mathbf{x}_k||^2}$
> $\mathbf{x}_{k+1} := \mathbf{v}/||\mathbf{v}||$
> $k = k + 1$
> *until* $|\lambda_k - \lambda_{k-1}| < \varepsilon$
> *Output*   Eigen pair $(1/\lambda_k, \mathbf{x}_k)$

Here lies the benefit from using the LU factorization since several iterations are often needed to obtain the eigenvalue with high accuracy.

Example 17 Use inverse iteration to compute the smallest eigenvalue of the matrix A used in Example 16.

Using the same starting vector as in Example 16 the first few estimates of the largest eigenvalue of the inverse of A are: $0.0110, -1.000, -2.500, -3.0000$ which are settling slowly. After another 22 iterations we have the converged value -3.7285 for the largest eigenvalue of A^{-1} so that the smallest eigenvalue of A is -0.2682 with the approximate associated eigenvector $[0.1487, -0.4074, 0.5587, -0.5600, 0.4076, -0.1413]'$.

Testing the residual of these estimates by computing $A\mathbf{v} - \lambda\mathbf{v}$ we get a maximum component of approximately 1.5×10^{-4}, which is consistent with the accuracy in the approximation of eigenvalue.

There is even more good news–the power method can be further modified to find other eigenvalues in addition to the largest and smallest in magnitude. Having some notion of the locations of those eigenvalues can help and the following theorem provides that insight.

Theorem 10 (Gerschgorin's Theorem) *Every eigenvalue of an $n \times n$ matrix A lies in one of the complex disks centered on the diagonal elements of A with radius equal to the sum of the off-diagonal elements in the corresponding row. These are the disks*

$$|z - a_{ii}| \le \sum_{j \ne i} |a_{ij}|.$$

Moreover, if any collection of m of these disks is disjoint from the rest, then exactly m eigenvalues (counting multiplicities) lie in the union of these disks.

Example 18 Apply Gerschgorin's theorem to the matrix of Examples 16 and 17.

The diagonal entries and the absolute sums of their off-diagonal row-elements give us the following set of centers and radii:

Center	1	2	3	4	5	6
Radius	20	20	21	23	26	30

These disks are plotted in Fig. 4.7. In this case Gerschgorin's Theorem gives us little new information since all the disks are contained in the largest one. Since the matrix A is symmetric we do know that all of its eigenvalues are real, so the theorem really only tells us that all eigenvalues lie in $[-24, 36]$. Since we already know that $\lambda_{\max} = 27.7230$ and $\lambda_{\min} = -0.2682$ to four decimals, we can reduce this interval to $[-24, 27.7230]$ but this does not really assist in predicting where the other eigenvalues lie.

Fig. 4.7 The Gerschgorin disks from Example 18

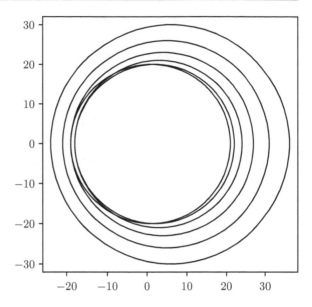

Despite the example, Gerschgorin's theorem can be useful. The matrix

$$B = \begin{bmatrix} 3 & 0 & 1 & -1 \\ 0 & 6 & 0 & 1 \\ 1 & 0 & -1 & 0 \\ -1 & 1 & 0 & 10 \end{bmatrix} \qquad (4.42)$$

is symmetric and so has real eigenvalues. Gerschgorin's theorem implies that these lie in the union of the intervals $[1, 5]$, $[5, 7]$, $[-2, 0]$, and $[8, 12]$. Note that since the last two of these are each disjoint from the others, we can conclude that one eigenvalue lies in each of the intervals $[-2, 0]$, and $[8, 12]$ while the remaining two lie in $[1, 5] \cup [5, 7] = [1, 7]$. The smallest and largest are -1.2409 and 10.3664 to four decimal places, respectively. There are two eigenvalues in $[1, 7]$ still to be found. If we can locate the one closest to the mid-point of this interval, then we could use the trace to obtain a very good approximation to the remaining one.

How then could we compute the eigenvalue closest to a particular value? Specifically suppose we seek the eigenvalue of a matrix A closest to some fixed μ.

Note that if λ is an eigenvalue of A with associated eigenvector $\mathbf{v}$, then $\lambda - \mu$ is an eigenvalue of $A - \mu I$ with the same eigenvector $\mathbf{v}$ since

$$(A - \mu I)\,\mathbf{v} = A\mathbf{v} - \mu\mathbf{v} = \lambda\mathbf{v} - \mu\mathbf{v} = (\lambda - \mu)\,\mathbf{v}$$

The smallest eigenvalue of $A - \mu I$ can be found using inverse iteration. Adding μ will then yield the desired eigenvalue of A closest to μ. This technique is called *origin shifting* since we "shift the origin" of the number line on which our eigenvalues lie.

Example 19 Find the eigenvalue of the matrix B given by (4.42) closest to 4.

Applying inverse iteration to the matrix

$$
C = B - 4I = \begin{bmatrix} -1 & 0 & 1 & -1 \\ 0 & 2 & 0 & 1 \\ 1 & 0 & -5 & 0 \\ -1 & 1 & 0 & 6 \end{bmatrix}
$$

gives the smallest eigenvalue as -0.907967 to six decimals. Adding 4 gives the closest eigenvalue of B to 4 as 3.092033.

The trace of B is $3 + 6 - 1 + 10 = 18$ and this must be the sum of the eigenvalues. We therefore conclude that the remaining eigenvalue must be very close to

$$
18 - (-1.2409) - 3.0920 - 10.3664 = 5.7825
$$

If this eigenvalue was sought to greater accuracy, we could use inverse iteration with an origin shift of, say, 6.

Although we have not often reported them, all the techniques based on the power method also provide the corresponding eigenvectors.

What about the remaining eigenvalues of our original example matrix from Example 16? Gerschgorin's theorem provided no new information. The matrix was

$$
A = \begin{bmatrix} 1 & 2 & 3 & 4 & 5 & 6 \\ 2 & 2 & 3 & 4 & 5 & 6 \\ 3 & 3 & 3 & 4 & 5 & 6 \\ 4 & 4 & 4 & 4 & 5 & 6 \\ 5 & 5 & 5 & 5 & 5 & 6 \\ 6 & 6 & 6 & 6 & 6 & 6 \end{bmatrix}
$$

and we have largest and smallest eigenvalues $\lambda_{\max} = 27.7230$ and $\lambda_{\min} = -0.2682$ to four decimals.

Example 20 Find the remaining eigenvalues of the matrix A above.

The sum of the eigenvalues is the trace of the matrix, 21, and their product is $\det A = -6$. Since

$$
\lambda_{\max} + \lambda_{\min} = 27.7230 - 0.2682 = 27.4548
$$

the sum of the remaining values must be approximately -6.5. Their product must be close to $-6/ (27.7230) (-0.2682) = 0.807$. Given the size of the product it is reasonable to suppose there is one close to -5 or -6 with three smaller negative ones.

Using an origin shift of -5 and inverse iteration on $A - (-5)I$ we get one eigen-value at -4.5729.

The final three have a sum close to -2 and a product close to $0.8/(-4.6) = -0.174$. Assuming all are negative, then one close to -1 and two small ones looks likely. Using an origin shift of -1, we get another eigenvalue of A at -1.0406.

Finally, we need two more eigenvalues with a sum close to -1 and a product around 0.17. Using an origin shift of -0.5 we get the fifth eigenvalue -0.5066. The trace now gives us the final eigenvalue as being very close to

$$21 - 27.7230 - (-4.5729) - (-1.0406) - (-0.5066) - (-0.2682) = -0.3347$$

What we have seen in this last example is that for a reasonably small matrix, which we can examine, then the combination of the power method with inverse iteration and origin shifts *can* be used to obtain the full eigen structure of a matrix all of whose eigenvalues are real. Of course this situation is ideal but perhaps not realistic. It may not be known in advance that the eigenvalues are all real. The possible sign patterns of the "missing" eigenvalues might be much more difficult to guess than in this example.

Luckily, efficient and accurate numerical techniques do exist to approximate eigenvalues of large linear systems. Those methods are often studied in detail in upper level Matrix Theory and Computations courses. The purpose of this section serves as an introduction to the subject.

Exercises

Consider the matrix

$$A = \begin{bmatrix} 5 & 5 & 5 & 5 & 5 \\ 4 & 4 & 4 & 4 & 5 \\ 3 & 3 & 3 & 4 & 5 \\ 2 & 2 & 3 & 4 & 5 \\ 1 & 2 & 3 & 4 & 5 \end{bmatrix}$$

which has entirely real eigenvalues. Find these eigenvalues each accurate to 5 decimal places.

1. Find the largest eigenvalue of A using the power method.
2. Find the smallest eigenvalue of A using inverse iteration. (Write a program for inverse iteration as you will be using it in the subsequent exercises.)
3. Given that there is positive eigenvalue close to 3, use inverse iteration with an origin shift to locate this eigenvalue.
4. Repeat Exercise 3 for the eigenvalue close to -1.
5. Use the trace and determinant to find a suitable origin shift and then another eigenvalue.
6. Use the trace to estimate the final eigenvalue. Use a suitable origin shift to compute this last one and to provide an accuracy check.

7. Modify your subroutine for the inverse power method so that it uses the LU decomposition. Compare the computation costs of the two approaches for finding the smallest eigenvalue of A.

8. One application of eigenvalues arises in the study of earthquake induced vibrations on multistory buildings. The free transverse oscillations satisfy a system of second order differential equations of the form

$$mv'' = kBv,$$

where here B is an $n \times n$ matrix if there are n floors in the building, m is the mass of each floor, and k denotes the stiffness of the columns supporting the floor. Here $v(t)$ is a vector that describes the deformation of the columns. B has the following tridiagonal structure

$$B = \begin{bmatrix} -2 & 1 & 0 & \cdots \\ 1 & -2 & 1 & 0 \\ 0 & & \ddots & \\ \vdots & & 1 & -2 \end{bmatrix}.$$

(a) Verify that if w is an eigenvector of B with eigenvalue λ then for $\omega = -\lambda(k/m)$ and for any constant α, we have that $v(t) = \alpha \cos(\omega t)w$ is a solution to the above ordinary differential equation.

(b) Use the inverse power method to find the smallest eigenvalue of B.

(c) Let $m = 1250, k = 10000$, and and compute a solution vector $v(t) = [v_1(t), v_2(t), v_3(t), v_4(t), v_5(t)]^T$ to the ODE above. Plot the parametric curves $v_k(t) + k$ for $k = 1, 2, 3, 4, 5$ where t ranges from 0 to 30. What do these curves describe? The following template will help organize your solution:

```
import numpy as np
from matplotlib import pyplot as plt

# Eigenvalue example for earthquake induced
# vibrations define the matrix
B = np.array([[-2,  1,  0,  0,  0],
              [ 1, -2,  1,  0,  0],
              [ 0,  1, -2,  1,  0],
              [ 0,  0,  1, -2,  1],
              [ 0,  0,  0,  1, -2]])
k = 1000
m = 1250
a = 0.075

# Find the smallest eigenvalue and accompanying
```

```
# eigenvector
# ...YOU DO THIS
# Call them lam and v

# Define omega
w = −1*lam *(k/m)

# Get ready to define your solutions
# YOU CREATE a vector t of 200 equally spaced
# points between 0 and 30
t =
# YOU Initialize your solution as a 5x200 array of
# zeros
Vk =

# YOU Define the solutions at each time step
# (THINK 5x200)
for i in range(200):
    Vk[:,i] =

# Plot the parametric curves
for i in range(5):
    plt.plot(t, Vk[i,:] + i)
plt.show()
```

9. We will consider a model for population growth that models only the female portion of a population and see how eigenvalues can help explain how the population evolves. Let L be the maximum age that can be attained by any female of the species. We divide the lifetime interval $[0, L]$ into n equally spaced subintervals to form n age classes as shown in the following table.

Age class	Age Interval
1	$[0, L/n)$
2	$[L/n, 2L/n)$
3	$[2L/n, 3L/n)$
$\vdots$	$\vdots$
$n-1$	$[(n-2)L/n, (n-1)L/n)$
n	$[(n-1)L/n, L)$

So for example, the age a of a female in class 3 must satisfy $2L/n \le a < 3L/n$. Next, suppose the vector $\mathbf{x}$, represents the female population at some particular point in time such that the ith component of $\mathbf{x}$ is the number of females in the ith age class. We also require the following assumptions to develop our model:

- The populations are measured at equally spaced time intervals, t_1, t_2, t_3, ..., and the vector $\mathbf{x}^j$ will represent the population at the end of the jth time interval when $t = t_j$.
- The time between measurements will be the same as the length of the age intervals used to classify the population. For example, if $L = 20$ and the number of age classes is five then the time between measurements is four years.
- The only factors that will be used to determine how the population changes are birth, death, and aging, so we will not consider factors like immigration and emigration. To take into account births and deaths, we introduce the following parameters:
- The birth coefficients a_i, $i = 1, 2, 3, \ldots, n$ denote the average number of females born to a single female while she is a member of the ith age class. For example, if $L = 20$ and $n = 5$, then a_3 would correspond to the average number of female babies born to a female in the third age class (females between the ages of 8 and 12).
- The survival coefficients, b_i, $i = 1, 2, 3, \ldots, n$ correspond to the fraction of females in age class i that will survive and pass to the $(i + 1)^{st}$ class. For example, suppose there are 100 females in age class 4 and that $b_4 = 3/4$ then we would predict that 75 females would be in class 5 the next time that the population is measured. Note that our assumption that no female lives longer than L years means that $b_n = 0$.

Given the above parameters, $\mathbf{x}^{k+1}$ can be calculated in terms of $\mathbf{x}^k$ with the linear model $A\mathbf{x}^k = \mathbf{x}^{k+1}$, where

$$
A = \begin{bmatrix}
a_1 & a_2 & \cdots & a_n \\
b_1 & 0 & \cdots & 0 \\
& & \ddots & \\
0 & & b_{n-1} & 0
\end{bmatrix}.
$$

Population Stability: There are three things that can happen to a population over time; the population (1.) increases with time, (2.) decreases with time, or (3.) the population stabilizes. In the case of (3.) we say that the population has reach steady state. Eigenvalues can help us predict when the above situations might occur. In particular, if A has a positive, dominant eigenvalue, then that eigenvalue can help determine the long-term behavior of the population.

(a) Explain how the structure of A is consistent with the definitions of the birth and survival coefficients and the way that the population advances in time.

(b) Suppose that a given population matrix A does have a dominant positive eigenvalue λ. Given that $A\mathbf{x}^k = \mathbf{x}^{k+1}$, suppose that for some m, that eventually $\mathbf{x}^m$ is the eigenvector of that corresponds to λ. How can we relate the size of λ to the three possible scenarios (think about $0 < \lambda < 1$, $\lambda = 1$, $\lambda > 1$).

Table 4.2 Population 1

Age class	a_i	b_i
1	0	1/2
2	1	3/4
3	2	4/5
4	4	1/2
5	2	–

Table 4.3 Population 2

Age class	a_i	b_i
1	0	1/8
2	1	1/4
3	3	1/2
4	4	1/4
5	1	–

Table 4.4 Population 3

Age class	a_i	b_i
1	0	1/2
2	1	1/2
3	1	1/2
4	1	1/2
5	2	–

(c) Consider the following sets of birth and death parameters (Tables 4.2, 4.3 and 4.4) of a population with $L = 20$ and $n = 5$. For each data set, model the population for a three choices of initial populations x^0 and explain what you see. Code up the power method for locating the dominant eigenvalue and corresponding eigenvector. Use it to analyze these three sets and show that the results are consistent with the theory you described above.

10. Consider the model in Exercise 9 above. Using whatever resources necessary, investigate some 'vital statistics', which give population information for various states and counties (actuaries work with this kind of data all the time!) Pick a 'population' and using real data that you find, create your own population model and analyze the long term behavior. If your eigenvalue locator doesn't work on the real problem, explain why not.

4.7 Conclusions and Connections: Linear Equations

What have you learned in this chapter? And where does it lead?

The basic problem of solving two (or more) linear equations in two (or more) unknowns is somewhat familiar. That situation was our starting point using simple examples to illustrate the basics of eliminating variables from one equation by using information from others.

That elementary approach is the basis for Gauss elimination. The need to automate the algorithm for computer application removes some of the choices that we have in the "hand solution" of small scale systems–"convenient" arithmetic for hand calculation is no longer a priority for the computer, of course.

Convenience may not be important but other arithmetic considerations certainly are. The effects of rounding error in a simplistic implementation of Gauss elimination can be severe–as we saw with a 3×3 example. Using pivoting, which is essentially equivalent to changing the order of the equations, can usually overcome this loss of precision.

A further modification, LU factorization, allows for subsequent systems with the same coefficient matrix to be solved much more efficiently. We saw this in practice for iterative refinement of the computed solution. The basic approach is still that of Gauss elimination but with the ability to do the manipulations on the matrix just once. For small systems that does not seem to offer much, but for large systems it can be significant. Frequently in large time-dependent computational models (such as weather systems modeling to take an extreme case) the same system needs to be solved for different data relevant to the next time-step. Changing the computational effort for subsequent data from proportional to n^3 to just proportional to n^2 (which is the effect of LU factorization) can result in order of magnitude savings in computational time and effort. If you want tomorrow's weather forecast today, rather than next week, such savings are vital.

The situation just referenced is not completely realistic because the large systems arising in big time-dependent models are often sparse, meaning lots of zeros in the coefficient matrix. In such situations it is common to use iterative methods rather than the so-called direct methods discussed up to now. The two fundamental approaches are Jacobi and Gauss–Seidel iterations. Here again, though, we have introduced a topic which is still actively researched. Large sparse linear systems play such an important role as a basic tool in major computational processes that seeking even modest improvements in speed, accuracy, storage requirements, memory handling, and inter-processor communication can pay large dividends.

The final situation in which we studied linear equations was within (linear) least squares approximation. It is worth reminding you that linear least squares does not mean fitting a straight line to data–though that can be an example. It refers to the problem being linear in the coefficients of the approximating function. Much of what was discussed related to finding least squares polynomial approximations. Yet again this is an introduction to a much bigger active research field. Fourier series can be obtained in this same way using linear combinations of trigonometric functions, for example. The Fast Fourier Transform, FFT, is one important example. It is arguably the most important workhorse of signal processing which in turn is used, often with special purpose improvements, in areas as diverse as hearing aid technology, radio and television transmissions in compressed data, security scanners, and medical imaging.

The final topic of the chapter was the eigenvalue problem. Eigenvalues, and their even more powerful cousins singular values, are vital tools for many purposes–signal processing is again an important one. They are the mathematical basis of many statistical data analysis techniques such as principal component analysis. As such eigenvalue problems are critically important to the whole field of data science or data analytics which has become one of the biggest growth areas for employment in recent years in industries from finance to health care to physical science and engineering.

4.8 Python's Linear Algebra Functions

4.8.1 Linear Equations

The fundamental Python (NumPy) function for linear equation solving is the numpy.linalg.solve function.

numpy.linalg.solve   This function solves a system of linear equations. The syntax is fairly straightforward. Consider a linear system

$$A\mathbf{x} = \mathbf{b}$$

This can be solved as

```
>>> x = np.linalg.solve(A, b)
```

LU factorization   The algorithm has been discussed in this chapter and we have used the Python routine in some examples. The syntax is:

```
>>> from scipy.linalg import lu
>>> P, L, U = lu(A)
```

This computes the LU factors of A as well as the permutation matrix P such that $A = PLU$ using partial pivoting. L in this case is strictly lower triangular and the row interchanges are separated into the permutation matrix P.
The variant

```
>>> L, U = lu(A, permute_l = True)
```

Computes the LU factors of A with permutations incorporated in the matrix L. In this case L is generally not strictly lower triangular due to row interchanges.

QR factorization   This factorization decomposes a matrix A into into a product of an orthonormal matrix Q and an upper triangular matrix R. When a matrix has mutually *ortho*gonal columns each *normal*ized to have norm 1, such a matrix is an *orthonormal matrix* . The QR factorization works for square as well as rectangular matrices. For an $m \times n$ matrix A, the orthonormal matrix Q is $m \times m$ and the upper "triangular" matrix R is $m \times n$.

The inverse of an orthonormal matrix is its own transpose which conveniently allows solving solving $Q\mathbf{y} = \mathbf{b}$ simply by setting $\mathbf{y} = Q^T\mathbf{b}$. Therefore, we can solve a linear system $A\mathbf{x} = \mathbf{b}$ through the following approach:

```
>>> from scipy.linalg import qr,
    solve_triangular
>>> Q, R = qr(A)
>>> x = solve_triangular(R, Q.T.dot(b))
```

scipy.linalg.solve_triangular is a function for solving linear equation systems which can exploit the fact that the matrix involved has upper triangular structure.

4.8.2 Linear Least Squares

numpy.polyfit   The polyfit function computes a discrete least squares fit of a polynomial to a set of data.

Consider a set of data points x_k $(k = 0, 1, \ldots, n)$ and corresponding function values f_k. We wish to fit a polynomial of degree $m < n$ to the set of data points. This results in the following linear system

$$a_0 + a_1 x_k + a_2 x_k^2 + \cdots + a_m x_k^m = f_k$$

for $k = 0, 1, \ldots, n$. We have n equations in m unknown coefficients. Solving a system of this kind in the least-squares sense corresponds to determining the values $a_0, a_1, \ldots, a_m$ that minimize the sum of the squared residuals

$$\sum_{k=0}^{n} \left| f_k - \left(a_0 + a_1 x_k + a_2 x_k^2 + \cdots + a_m x_k^m\right) \right|$$

The vector of coefficients returned by numpy.polyfit starts from the the highest powers of the polynomial. As an example, consider the data

```
>>> x = np.arange(-1, 1.1, 0.1)
>>> y = np.cos(np.pi * x)
```

A third-degree polynomial can be fitted to the data using the following command

```
>>> p = numpy.polyfit(x, y, 3)
```

The least-squares cubic approximation is given by the following coefficients

```
[ -7.7208e-15  -2.1039e+00  5.2187e-15  7.2382e-01]
```

corresponding to the "cubic"

$$-0x^3 - 2.1039x^2 + 0x + 0.7238$$

if we neglect the third- and first-degree coefficients which are essentially zero due to their small magnitude.

4.8.3 Eigenvalues

numpy.linalg.eig The NumPy function eig is the basic routine for computing eigenvalues and eigenvectors. It returns all eigenvalues and eigenvectors of a square matrix.
Running

```
>>> W, V = np.linalg.eig(A)
```

computes all eigenvalues of A returned as the vector W. The matrix V contains the corresponding eigenvectors.
scipy.linalg.eig The similar SciPy function eig can additionally be used for solving the *generalized eigenvalue problem*

$$A\mathbf{x} = \lambda B\mathbf{x}$$

numpy.linalg.svd The *Singular Value Decomposition*, abbreviated SVD, generalizes the concept of eigenvalue decomposition to rectangular matrices instead of just square matrices. The SVD factors rectangular matrices in the form

$$A = USV^\mathrm{T}$$

The matrices U, V are orthonormal and S is a diagonal matrix containing the *singular values* of A along its diagonal. The singular values are related to eigenvalues in the sense that they are the square-roots of the eigenvalues of the matrix $A^\mathrm{T}A$.
numpy.linalg.matrix_rank We often need to be able to compute the *rank* of a matrix. The rank is the number of linearly independent rows (or columns) of a matrix. The rank also corresponds to the number of nonzero singular values of a rectangular matrix (non-zero eigenvalues in case of a square matrix). NumPy's function matrix_rank uses SVD to compute the rank of a matrix. It is necessary to use a certain tolerance in determining whether individual singular values should be considered "zero". This is due to roundoff errors, as discussed in Chap. 2. Python uses a default value proportional to the largest singular value, to the largest dimension of the matrix, and to the machine precision of the data type in use for storing the matrix.
The syntax is as follows:

```
>>> r = np.linalg.matrix_rank(A)
```

for default tolerance. The tolerance (here 10^{-8} for example) can be specified using

```
>>> r = np.linalg.matrix_rank(A, tol = 1e-8)
```

4.8.4 Basic Linear Algebra Functions

There are numerous examples standard linear algebra "arithmetic" throughout this chapter. Here we also mention a small selection of other useful functions.

numpy.dot   computes the product of two vectors, two matrices, or a vector and a matrix. Depending on the dimensions of the vectors and matrices involved, the product may correspond to an inner or outer product, or in general just matrix-vector product. If **a**, **b** are two one-dimensional arrays (vectors), the following

```
>>> np.dot(a, b)
```

or equivalently

```
>>> a.dot(b)
```

corresponds to the inner product between the vectors:

$$\mathbf{a} \cdot \mathbf{b} = \sum a_k b_k$$

For example,

```
>>> a = np.arange(1,4)
>>> b = np.arange(6,9)
>>> np.dot(a, b)
```

returns the result 44. Note here that NumPy's arange function returns an array of integers from the first argument up to, but *excluding* the second argument. This is the standard convention in Python whenever specifying ranges of integers. They can be thought of as half-open intervals; closed at the low end and open at the high end.

numpy.linalg.norm   computes the Euclidean norm of a vector (or a matrix)

$$\|\mathbf{a}\| = \sqrt{\mathbf{a} \cdot \mathbf{a}}$$

Any p-norm for p a positive integer can also be obtained using np.linalg.norm, as well as $\|\cdot\|_{-\infty}$, $\|\cdot\|_{\infty}$, the Frobenius, and the nuclear norm. For example with **b** as above, we get

```
>>> np.linalg.norm(b)
12.206555615733702

>>> np.linalg.norm(b,1)
21.0

>>> np.linalg.norm(b,np.inf)
8.0
```

The default is the Euclidean or 2-norm.

Iterative Solution of Nonlinear Equations

5

5.1 Introduction

We have just seen that although linear functions have been familiar to us since grade school, large linear systems can be challenging to solve numerically. Moreover, even though linear models are used frequently in understanding relationships and making predictions, one could argue that the world around us is truly nonlinear (and possibly even unpredictable!)

As an example, consider pouring a hot cup of coffee (or tea, or your favorite hot beverage) and placing it on the kitchen counter. What things affect the temperature of the coffee? What happens to the *rate* at which the coffee is cooling over time? Could you make a sketch of the temperature of the coffee over time? Does it look linear? Newton's Law of Cooling is one way to approach this scenario with a well-known mathematical model.

Newton's law of cooling states that the rate of change of the temperature of an object is proportional to the difference between the object and the ambient environment. In ideal circumstances, the object's temperature can be modeled over time as

$$T(t) = T_E + (T_0 - T_E)e^{-kt}$$

Here T_E is the surrounding temperature of the environment and T_0 is the initial temperature. The parameter k describes how quickly the object cools down and depends on the object itself. This is a nonlinear model for the temperature. If you wanted to find the time when then coffee was your ideal drinking temperature, say $\hat{T}$, then you would need to solve the nonlinear equation of $\hat{T} - T(t) = 0$. We'll revisit this model later.

For now consider general problems of the form

$$f(x) = 0. \tag{5.1}$$

© Springer International Publishing AG, part of Springer Nature 2018
P. R. Turner et al., *Applied Scientific Computing*, Texts in Computer Science,
https://doi.org/10.1007/978-3-319-89575-8_5

This problem is surprisingly more challenging than it might seem and there are few functions in which we can directly find the solutions (or roots) of such an equation. To this end, the methods discussed here compute a *sequence* of approximate solutions which we hope *converges* to this solution. This idea was briefly presented in the previous chapter in the context of the Jacobi and Gauss-Seidel methods for linear systems. We provide more details with a brief listing of the basic facts included at the end of this introductory section for reference purposes. However, a deeper treatment of the convergence of sequences in general can be found in standard texts on calculus or elementary real analysis.

Iterative methods start with an initial estimate, or guess, of the solution and according to some simple rule, generate a sequence of estimates, or *iterates*, which we hope gets closer to the solution of the original problem. In addition to an initial iterate, iterative methods require a prescribed stopping criteria, which can impact the quality of the final solution. We shall also discuss the convergence properties of the methods. That it is, it is critical to know under what conditions a certain method will work and how quickly. The primary motivation for these theoretical aspects is to be able to make the right choices when selecting a solver and setting the algorithmic parameters, thereby avoiding unnecessary computations to obtain a level of precision in a computed answer. These following ideas will help in the study and implementation of iterative solvers.

5.1.1 Summary of Convergence of Sequences

Definition 11 A sequence of real numbers (a_n) converges to a limit L, written

$$a_n \to L \text{ as } n \to \infty$$

or

$$\lim_{n \to \infty} a_n = L$$

if for every (small) $\varepsilon > 0$ there exists a number N such that

$$|a_n - L| < \varepsilon$$

for every $n > N$.

The condition $|a_n - L| < \varepsilon$ can often be usefully rewritten in the form

$$L - \varepsilon < a_n < L + \varepsilon$$

- $1/n^p \to 0$ as $n \to \infty$ for every positive power p
- $x^n \to 0$ as $n \to \infty$ for any $|x| < 1$, (x^n) diverges if $|x| > 1$ or if $x = -1$
- If $a_n \to a$, $b_n \to b$ then

1. $a_n \pm b_n \to a \pm b$
2. $a_n b_n \to ab$ (In particular, $ca_n \to ca$ for any constant c.)
3. $a_n/b_n \to a/b$ provided $b_n, b \neq 0$ (In particular, $1/b_n \to 1/b$.)

- **Squeeze Theorem**: If $a_n \leq c_n \leq b_n$ for every n and $\lim a_n = \lim b_n = L$, then $\lim c_n = L$, also.
- If (a_n) is an *increasing* sequence (that is, $a_{n+1} \geq a_n$ for every n) which is *bounded above* (there exists a number $M \geq a_n$ for every n) then (a_n) converges. A similar statement applies to *decreasing* sequences.
- If $a_n \to a$ and f is continuous at a, then $f(a_n) \to f(a)$ (provided the terms of the sequence are in the domain of f).

5.2 The Bisection Method

Many ideas in this section that lead to numerical methods for solving $f(x) = 0$ are derived from ideas you saw in Calculus. One of the simplest approaches is the method of *bisection* which is based on the *Intermediate Value Theorem*. Suppose that f is continuous on an interval $[a, b]$ and that $f(a) f(b) < 0$ (so that the function is negative at one end-point and positive at the other) then, by the intermediate value theorem, there is a solution of Eq. (5.1) between a and b.

The basic situation and the first two iterations of the bisection method are illustrated in Fig. 5.1. At each stage we set m to be the midpoint of the current interval $[a, b]$. If $f(a)$ and $f(m)$ have opposite signs then the solution must lie between a and m. If they have the same sign then it must lie between m and b. One of the endpoints can then be replaced by that midpoint and the entire process is repeated on the new smaller interval.

In Fig. 5.1, we see that for the original interval $[0, 1.1]$, $m = 0.55$ and $f(a)$, $f(m)$ have the same sign. The endpoint a is replaced by m, which is to say m becomes

Fig. 5.1 Illustration of the bisection method for an arbitrary function

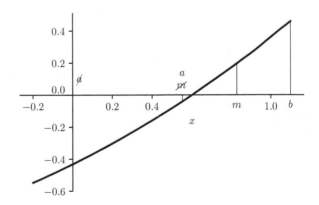

the new a, and a new m is computed. It is the midpoint of $[0.55, 1.1]$, $m = 0.825$ as shown. This time $f(a)$ and $f(m)$ have opposite signs, $f(a) f(m) < 0$, and so b would be replaced by m.

The process can be continued until an approximate solution has been found to a desired accuracy, or *tolerance*. One possible stopping criteria could be when the length of the interval $b - a$ is smaller than a specified tolerance.

The bisection method is summarized below.

Algorithm 12 Bisection method

Input	Equation $f(x) = 0$
	Interval $[a, b]$ such that $f(a) f(b) < 0$
	Tolerance ε
Repeat	$m := (a + b)/2$
	if $f(a) f(m) \leq 0$ then $b := m$ else $a := m$
until	$b - a < \varepsilon$
Output	Solution lies in the interval $[a, b]$ which has length less than ε

Example 1 Use the bisection method to solve

$$x - \cos x - 1 = 0 \tag{5.2}$$

with a tolerance of 0.1.

With $f(x) = x - \cos x - 1$, we see easily that

$$f(0) = -2 < 0, \quad f(\pi/2) = \pi/2 > 0.$$

Since f is continuous, the intermediate value theorem shows that there is a solution of (5.2) in $[0, \pi/2]$. We set $a = 0, b = \pi/2$. For the first iteration, $m = \pi/4$, $f(\pi/4) = \pi/4 - \cos \pi/4 - 1 = -0.9217 < 0$. So the midpoint, m now becomes the left endpoint of the search interval, i.e. the solution must be in $[\pi/4, \pi/2]$ so we set $a = \pi/4$.

As the algorithm progresses, it can be useful to organize all the necessary information in a table. For the second iteration $m = 3\pi/8$, $f(3\pi/8) = 3\pi/8 - \cos \pi/8 - 1 = -0.2046 < 0$. So the solution lies in $[3\pi/8, \pi/2]$ (i.e. so we set $a = 3\pi/8$ for the next iteration.

At this point the process stops because $b - a \approx 0.09 < 0.1$ and the solution is 1.2836 with error less than 0.05. Note that the solution is the midpoint of the final interval and so is less than $\varepsilon/2$ away from either end of the interval.

it	m	f(a)	f(m)	f(b)	a	b	b − a
1	0.7854	−2.0000	−0.9217	0.5708	0.7854	1.5708	0.7854
2	1.1781	−0.9217	−0.2046	0.5708	1.1781	1.5708	0.3927
3	1.3744	−0.2046	0.1794	0.5708	1.1781	1.3744	0.1963
4	1.2763	−0.2046	−0.0140	0.1794	1.2763	1.3744	0.0982

Clearly if a highly accurate solution is required, convergence could be slow. However the description in Algorithm 12 lends itself to a computer code in a straightforward way, as seen in the next example.

Example 2 Solve equation (5.2) used in Example 1 using the bisection method in Python. Start with the interval $[0, n/2]$ and obtain the solution with an error less than 10^{-5}.

First we define a function eq1 for the equation as follows

```python
from math import cos, π

def eq1(x):
    return x − cos(x) − 1
```

Then the following instructions in the Python command window will achieve the desired result:

```python
a = 0
b = π /2
tol = 1e−5
fa = eq1(a)
fb = eq1(b)
while abs(b − a) > 2 * tol:
    m = (a + b) / 2
    fm = eq1(m)
    if fa * fm <= 0:
        b=m
    else:
        a=m
m = (a + b) / 2

>>> print(m)
1.283432589901804
```

Note that we could use 2*tol in the test for the while statement because of the final step after the completion of the loop which guarantees that this final solution has the required accuracy. Also we used abs(b-a) to avoid any difficulty if the starting values were reversed.

Of course, the trailing digits in the final answer do not imply that level of precision in the solution – we have only ensured that the true solution is within 10^{-5} of this value.

It should be fairly apparent that the commands used here correspond to those of Algorithm 12. One key difference between that algorithm and the Python commands above is that the latter depend critically on the *name* of the function file whereas in the algorithm the function is one of the inputs. This can be achieved by defining a function for the bisection method:

Program   Python function for the bisection method, Algorithm 12

```python
def bisect(fcn, a, b, tol):
    """
    Solve the equation fcn(x)=0 in the interval
    [a,b] to accuracy 'tol' using the bisection
    method. 'fcn' is a function.

    """

    fa = fcn(a)
    fb = fcn(b)
    while abs(b - a) > 2 * tol:
        m = (a + b) / 2
        fm = fcn(m)
        if fa * fm <= 0:
            b = m
        else:
            a = m
    return (a + b) / 2
```

To see the use of this bisection algorithm function, we resolve the same equation to greater accuracy with the single command:

```python
>>> s = bisect(eq1, 0, π/2, 1e-6)
>>> print(s)
1.283428844831521
```

The program above will work provided the function is continuous and provided that the initial values fa and fb have opposite signs. It would be easy to build in a check that this condition is satisfied by inserting the lines

Fig. 5.2 Illustration of the function from Example 3

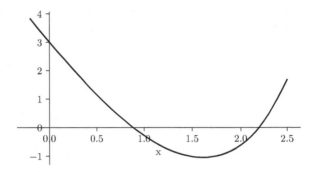

```
if fa * fb > 0:
    print('Two endpoints have same sign')
    return
```

immediately before the start of the while loop.

Example 3 Use the bisection method to find the positive solutions of the equation $f(x) = e^x - 5x + 2$ by the bisection method.

We begin by defining the function involved in the equation:

```
def eq2(x): return exp(x) - 5 * x + 2
```

First we shall try the initial interval [0.1, 1]. With this function defined as eq2 the command

```
>>> s = bisect(eq2, 0, 2, 1e-6)
>>> print(s)
0.8842172622680664
```

finds the first of two roots.

The second root in the interval [2, 3] can be found as follows:

```
>>> s = bisect(eq2, 2, 3, 1e-6)
>>> print(s)
2.193741798400879
```

We can convince ourselves that the solutions seem accurate from the plot of the function in Fig. 5.2.

We conclude this section with an outline of the proof that this method indeed converges for any continuous function with $f(a) f(b) < 0$. (Refer to the convergence definitions of sequences at the beginning of this chapter as needed!)

Let the sequences of endpoints be denoted by (a_n) and (b_n) and suppose $a_0 < b_0$. Now

$$a_{n+1} = \begin{cases} a_n & \text{if } f(a_n) f(m) < 0 \\ m = \dfrac{a_n + b_n}{2} & \text{if } f(a_n) f(m) \geq 0 \end{cases}$$

It follows that (a_n) is an increasing sequence while (b_n) is decreasing.

Moreover $a_n < b_n \leq b_0$ for every n. That is (a_n) is increasing *and* bounded above. Therefore (a_n) converges to some limit, L, say. Similarly (b_n) is decreasing and bounded below (by a_0) and so also converges, to M, say.

The bisection algorithm is constructed so that

$$b_n - a_n = \frac{b_{n-1} - a_{n-1}}{2}$$

which implies that

$$b_n - a_n = \frac{b_0 - a_0}{2^n}$$

and therefore $b_n - a_n \to 0$. It follows that $L = M$; that is the two sequences have the same limit.

Finally, because f is a continuous function,

$$f(a_n) f(b_n) \to [f(L)]^2 \geq 0$$

but the algorithm is designed to ensure that $f(a_n) f(b_n) \leq 0$ for every n. The only possibility is that $f(L) = 0$ which is to say this limit is a solution of the equation $f(x) = 0$.

Exercises

1. Show that the equation

   $$3x^3 - 5x^2 - 4x + 4 = 0$$

   has a root in the interval $[0, 1]$. Use the bisection method to obtain an interval of length less than $1/8$ containing this solution. How many iterations would be needed to obtain this solution with an error smaller than 10^{-6}?

2. Show that the equation $e^x - 3x - 1 = 0$ has only one positive solution, and that it lies in the interval $[1, 3]$. Use the bisection method to find the positive solutions of this equation with error less than 10^{-5}.

3. Show that the equation

   $$e^x - 100x^2 = 0$$

   has exactly three solutions. Obtain intervals of length less than 0.1 containing them.

4. Consider the coffee cooling scenario above. Use $T_E = 22\,°C$, $T_0 = 90\,°C$ and suppose your ideal temperature to drink the coffee is $\hat{T} = 80\,°C$. Find the time that the coffee reaches that temperature by solving $\hat{T} - T(t) = 0$ if $k = 0.04$ using the bisection method.

5.3 Fixed Point Iteration

Although the bisection method provides a simple and reliable technique for solving an equation, if the function f is more complicated (perhaps its values must be obtained from the numerical solution of a differential equation) or if the task is to be performed repeatedly for different values of some parameters, then a more efficient technique is needed.

Fixed point iteration applies to problems in the special case that $f(x) = 0$ can be rewritten in the form

$$x = g(x). \qquad (5.3)$$

This new rearrangement can be used to define an iterative process as follows. Given an initial iterate x_0, or estimate of the solution, then a sequence can be defined by (x_n) with

$$x_n = g(x_{n-1}) \qquad n = 1, 2, \ldots . \qquad (5.4)$$

Provided that g is continuous, we see that, *if* this iterative sequence converges then the terms get closer and closer together. Eventually we obtain, to any required accuracy,

$$x_n \approx x_{n-1}$$

which is to say

$$x_{n-1} \approx g(x_{n-1})$$

so that x_n is, approximately, a solution of Eq. (5.3).

Example 4 We again consider (5.2), $x - \cos x - 1 = 0$, which we can obviously rewrite as

$$x = \cos x + 1$$

The *iteration function* is then $g(x) = \cos x + 1$. Consider the initial guess $x_0 = 1.374$ (which is roughly the second midpoint in the bisection iteration in Example 2) The algorithm would give the following sequence

$$x_1 = \cos 1.374 + 1 = 1.1955$$
$$x_2 = \cos 1.1955 + 1, 8 = 1.3665$$
$$x_3 = \cos 1.3665 + 1, 5 = 1.2029$$
$$\vdots$$
$$x_9 = 1.2212$$
$$x_{10} = 1.3425$$
$$x_{11} = 1.2263$$
$$x_{12} = 1.3377$$
$$x_{13} = 1.2310$$

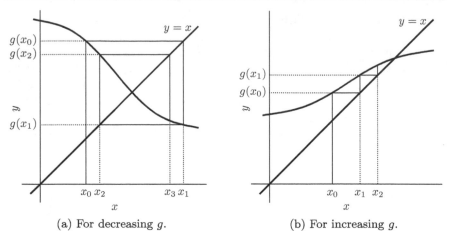

(a) For decreasing g.   (b) For increasing g.

Fig. 5.3 Illustration of converging function iterations

which appears to be converging to the same solution obtained with the bisection method (although nothing has been formally proven about convergence).

Given a problem $f(x) = 0$, there may be multiple ways to reformulate it as a fixed point iteration. For example, the same equation could have been rewritten as $x = \cos^{-1}(x - 1)$. Using the starting value $x_0 = 1.374$, the first few iterations yield the sequence

$$1.1875, \; 1.3822, \; 1.1786, \; 1.3912, \; 1.1688, \; 1.4011, \; 1.1580, \; 1.4121$$
$$1.1460, \; 1.4242, \; 1.1327, \; 1.4377, \; 1.1177, \; 1.4528, \; 1.1009, \; 1.4697$$

So the bad news is that after 18 iterations, the iteration is moving away from the solution in both directions and is eventually not defined.

There is also good news; it is possible to determine in advance if a particular re-formulation will converge *and* obtain a desired accuracy in the approximate solution. A graphical illustration of the process of function iteration helps to explain this. Solving an equation of the form $x = g(x)$ is equivalent to finding the intersection of the graphs $y = x$ and $y = g(x)$. Figure 5.3 illustrates two convergent function iterations. In each case, x_n is the point at which the vertical line $x = x_{n-1}$ meets the curve $y = g(x)$. The horizontal line at this height, $y = g(x_{n-1})$ meets the graph of $y = x$ at $x = x_n = g(x_{n-1})$. The process can be continued indefinitely.

In the first case, where g is a decreasing function, the iterates converge in an oscillatory manner to the solution – giving a web-like picture. This can be useful in determining the accuracy of the computed solution. In the second case, the iterates converge monotonically – either ascending or descending a staircase.

In Fig. 5.3, we see that function iteration can converge independent of whether the iteration function is increasing or decreasing near the solution. What turns out to be critical is *how fast* the function changes.

In the following analysis, assume that (5.3) is a rearrangement of (5.1) with g at least twice differentiable and the solution of this equation is denoted by s. Let (e_n) be the sequence of errors in the iterates given by

$$e_n = x_n - s$$

for each n.

The fact that $g(s) = s$ along with the Taylor series expansion of g about the solution gives

$$
\begin{aligned}
e_{n+1} = x_{n+1} - s &= g(x_n) - g(s) \\
&= \left[g(s) + (x_n - s) g'(s) + \frac{(x_n - s)^2}{2!} g''(s) + \cdots \right] - g(s) \\
&= e_n g'(s) + \frac{e_n^2}{2!} g''(s) + \cdots
\end{aligned}
\tag{5.5}
$$

If $|e_n|$ is small enough that higher-order terms can be neglected, we get

$$e_{n+1} \approx e_n g'(s) \tag{5.6}$$

which implies that the error will be reduced if $|g'(s)| < 1$. Keep in mind, we do not know s in advance, but the following theorem establishes that a sufficient condition for convergence of an iteration. The results is that convergence is guaranteed if $|g'(x)| < 1$ throughout an interval containing the solution. The condition that g is twice differentiable was used to simplify the analysis above. We see in the next theorem that only the first derivative is strictly needed.

Theorem 13 *Suppose that g is differentiable on $[a, b]$ and that*
(i) $g([a, b]) \subseteq [a, b]$ *(That is, $g(x) \in [a, b]$ for every $x \in [a, b]$), and*
(ii) $|g'(x)| \leq K < 1$ *for all $x \in [a, b]$.*
Then the equation $x = g(x)$ has a unique solution in the interval $[a, b]$ and the iterative sequence defined by

$$x_0 \in [a, b]; \quad x_n = g(x_{n-1}), \ n = 1, 2, \ldots$$

converges to this solution.

Proof Firstly, since $g(a), g(b) \in [a, b]$, it follows that

$$a - g(a) \leq 0 \leq b - g(b)$$

The intermediate value theorem implies that there exists $s \in [a, b]$ such that $s = g(s)$. The fact that s is the only such solution follows from the mean value theorem.

Suppose that $t = g(t)$ for some $t \in [a, b]$. By the mean value theorem, $g(s) - g(t) = (s - t) g'(\xi)$ for some ξ between s and t. But, $g(s) = s$ and $g(t) = t$ so that

$$s - t = (s - t) g'(\xi)$$

or

$$(s - t) \left[1 - g'(\xi) \right] = 0$$

By condition (ii), $\left| g'(\xi) \right| < 1$ which implies $s - t = 0$; proving that the solution is unique.

The convergence of the iteration is also established by appealing to the mean value theorem. By condition (i), we know that $x_n \in [a, b]$ for every n since $x_0 \in [a, b]$. Then, for some $\xi_n \in [a, b]$,

$$\begin{aligned} |e_{n+1}| = |g(x_n) - g(s)| &= |x_n - s| \cdot \left| g'(\xi_n) \right| \\ &= |e_n| \cdot \left| g'(\xi_n) \right| \leq K |e_n| \end{aligned}$$

In turn we obtain $|e_{n+1}| \leq K^{n+1} |e_0| \to 0$ which implies $x_n \to s$. ∎

Under the same conditions, if we choose $x_0 = (a + b)/2$, then $|e_0| = |x_0 - s| \leq (b - a)/2$ so that this result also provides bounds for the errors. In the next examples we see how to use this result to determine in advance whether an iteration will converge, and to estimate the accuracy of our results.

Example 5 Consider the equation of Example 3, $e^x - 5x + 2 = 0$ which we already know has a solution in $[2, 3]$. Three possible rearrangements are

$$x = \frac{e^x + 2}{5}, \qquad\qquad\qquad (i)$$
$$x = e^x - 4x + 2, \text{ and} \qquad\qquad (ii)$$
$$x = \ln(5x - 2) \qquad\qquad\qquad (iii)$$

We begin by examining the first few iterations using each of these rearrangements with the iteration functions defined as Python functions. In each case we used the initial guess $x_1 = 2.2$.

The Python commands used for the first of these were as follows. The modifications required for the other cases should be apparent.

```
x = [2.2]
for k in range(1,15):
    x.append(iter1(x[k - 1]))
```

where the function iter1 is defined as

```
from numpy import exp

def iter1(x):
    return (exp(x) + 2) / 5
```

Note the use of NumPy's exp rather than exp from the Python standard library's math module is necessary as the latter cannot handle very large function arguments. For large arguments that result in an overflow the above code will output the following warning

```
RuntimeWarning: overflow encountered in exp
    return (exp(x) + 2) / 5
```

The results for the three rearrangements are:

Iteration number	Rearrangement 1	Rearrangement 2	Rearrangement 3
1	2.2	2.2	2.2
2	2.205	2.22501	2.19722
3	2.21406	2.35355	2.19568
4	2.23055	3.10868	2.19482
5	2.261	11.9568	2.19434
6	2.31853	155831	2.19408
7	2.43215	inf	2.19393
8	2.67667	nan	2.19385
9	3.30733	nan	2.1938
10	5.86244	nan	2.19377
11	70.7164	nan	2.19376
12	$1.02989 \cdot 10^{30}$	nan	2.19375
13	inf	nan	2.19375
14	inf	nan	2.19374

The results imply that the first two rearrangements are not converging, while the third *appears* to be settling down, slowly. Theorem 13 helps explain why.

(*i*)   For the first rearrangement, $g(x) = \dfrac{e^x + 2}{5}$, giving $g'(x) = e^x/5 > 1$ for every $x > \ln 5 \approx 1.6$. The theorem indicates this iteration should not converge for the solution in $[2, 3]$. However, we see that $0 < g'(x) < 1$ for all $x < \ln 5$. Also, if $0 < x < \ln 5 < 2$, then $g(x) \in (0.06, 0.943)$. It follows that the conditions of the theorem are satisfied on this interval, and therefore that this iteration will converge to this solution. With $x_1 = 0.9$, the first few iterations yield:

$$0.8919, 0.8919, 0.8880, 0.8860, 0.8851, 0.8846, 0.8844, 0.8843, 0.8843, 0.8842, 0.8842$$

which appears to be settling down and $f(0.8842) = e^{0.8842} - 5(0.8842) + 2 \approx -6.8670e - 06$ (which we want to be close to zero and indeed is getting closer).

(*ii*)   This time, $g(x) = e^x - 4x + 2$, so that $g'(x) = e^x - 4$. This means that $g'(x) > 1$ for every $x > \ln 5$. and the iteration should not be expected to converge to the solution in $[2, 3]$. Unfortunately for this rearrangement, you can verify that

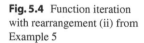

Fig. 5.4 Function iteration
with rearrangement (ii) from
Example 5

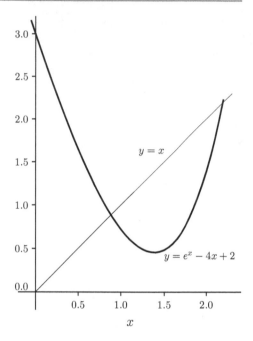

condition (i) of Theorem 13 is not satisfied on either $[0, 2]$ or $[2, 3]$ so convergence
will not be achieved. The rearrangement is illustrated in Fig. 5.4.

(iii) Using the third rearrangement, $g(x) = \ln(5x - 2)$, we obtain $g'(x) =$
$5/(5x - 2)$ which lies between 0.3846 and 0.6250 for $x \in [2, 3]$. It follows that g is
increasing, and, since $g(2) = 2.0794$ $g(3) = 2.5649$, it follows that the conditions
of the theorem are satisfied and so convergence is established for any starting point
in $[2, 3]$.

The implementation of function iteration in is straightforward since no additional
subroutines are needed beyond the iteration function itself. However, the user must
specify a stopping criteria. In the following implementation, the iteration is stopped
when either three successive iterates agree to within the tolerance (10^{-4} in this case),
or some maximum number of iterations (50) has been performed.

```
maxits = 50
tol = 1e−4
x0 = 1
x1 = 2
x2 = 1.5

for iteration in range(1,maxits):
    x0 = x1
    x1 = x2
    x2 = iter3(x1)
```

```
if  (abs(x1 − x0) < tol)  and  (abs(x2 − x0) <
    tol)  and  (abs(x2 − x1) < tol):
    break
```

```
>>> print(iteration)
18
```

```
>>> print(x2)
2.19370682
```

Note that in this implementation, we do not store the full array of iterates–only the last three. The initial values x0=1; x1=2; are essentially arbitrary, we just need to ensure that the iterative while loop starts. Here, the values of x0, x1 and x2 are continually overwritten as better estimates of the solution are computed. In this case, 15 iterations were enough to meet the prescribed accuracy. (It is common to test for convergence by examining just the last two iterates, but in the absence of other information this test is safer – even if a little too conservative.)

Although we have proved convergence, we can look closer at the accuracy as well by looking at the Taylor expansion of the error in Eq. (5.5) which shows that

$$e_{n+1} = e_n g'(s) + \frac{e_n^2}{2!} g''(s) + \cdots$$

Note that in Example 5, rearrangement (i) converged to the smaller solution for appropriate starting points. The solution is less than 0.9, and therefore $0 < g'(s) < g'(0.9) < 0.5$ so that as the iterates converge the errors satisfy

$$e_{n+1} < \frac{e_n}{2}$$

meaning each iterate reduces the error by at least this factor, as we saw in the numerical results. Sometimes we can do much better, however. If we obtain a rearrangement such that $g'(s) = 0$ then subsequent errors will be proportional to the squares of the prior ones. In particular if we also have $g''(s) < 2$ then errors satisfy

$$e_{n+1} < e_n^2$$

which will decrease much more rapidly.

This type of convergence is called *quadratic*, or *second-order convergence*. This implies that the exponent field in the error would double as the iteration progresses; for example if $e_n \approx 10^{-1}$ then the sequence of errors would approximately behave like 10^{-1}, 10^{-2}, 10^{-4}, 10^{-8}. Theorem 13 can be extended to provide error estimates–although possibly not tight error bounds, which may result in wasted computations. Instead of developing this analysis, we will focus on methods with faster convergence properties. In the next section, we shall develop a general method for obtaining quadratic convergence.

Exercises

1. The equation
$$3x^3 - 5x^2 - 4x + 4 = 0$$
   has a solution near $x = 0.7$. (See Exercise 1 in Sect. 5.2.) Carry out the first five
   iterations for each of the rearrangements

   (i) $x = \dfrac{5}{3} + \dfrac{4}{3x} - \dfrac{4}{3x^2}$,   and (ii) $x = 1 + \dfrac{3x^3 - 5x^2}{4}$

   starting with $x_0 = 0.7$.
2. Which of the iterations in Exercise 1 will converge to the solution near 0.7? Prove
   your assertion using Theorem 13. Find this solution using a tolerance of 10^{-6}.
3. Find a convergent rearrangement for the solution of the equation $e^x - 3x - 1 = 0$
   in $[1, 3]$. Use it to locate this solution using the tolerance 10^{-6}.
4. Intervals containing the three solutions of $e^x - 100x^2 = 0$ were found in Exercise 3 of Sect. 5.2. Each of the following rearrangements yields a convergent
   iteration for one of these solutions. Verify that they are all rearrangements of the
   original equation, and determine which will converge to which solution. Use the
   appropriate iterations to locate the solutions with tolerance 10^{-6}.

   (i) $x = \dfrac{\exp(x/2)}{10}$

   (ii) $x = 2(\ln x + \ln 10)$

   (iii) $x = \dfrac{-\exp(x/2)}{10}$

5.4 Newton's Method

Newton's method is one of the most widely used nonlinear solvers and it can be
derived using basic ideas from Calculus and the notion of approximating a nonlinear
function with a linear one. Moreover, the convergence of the iterative schemes in
Example 5 implied that rapid (quadratic) convergence was achieved when $g'(s) = 0$.
This is the motivation behind Newton's, or the Newton–Raphson, method. The
iteration can be derived in terms of the first-order Taylor polynomial approximation,
but essentially a local linear model using the simple tangent line to $f(x)$ is the
underlying idea.

   To see this, consider the original nonlinear problem (5.1) $f(x) = 0$. The first-
order Taylor expansion of f about a point x_0 is

$$f(x) \approx f(x_0) + (x - x_0) f'(x_0),$$

which looks a lot like the equation of the tangent line to $f(x)$ at x_0. If the
point x_0 is close to the required solution s, then we expect that setting the right-
hand side to zero should give a better approximation of this solution. Solving

$f(x_0) + (x - x_0) f'(x_0) = 0$ (which is equivalent to finding where the tangent line crosses the x-axis) provides an expression for the next approximation

$$x_1 = x_0 - \frac{f(x_0)}{f'(x_0)}$$

This process would then be repeated to generate a sequence of tangent lines, and roots of those tangent lines to get a general form of the *Newton iteration* formula

$$x_{n+1} = x_n - \frac{f(x_n)}{f'(x_n)} \qquad (n = 0, 1, 2, \ldots) \qquad (5.7)$$

Newton's method therefore uses the iteration function

$$g(x) = x - \frac{f(x)}{f'(x)}$$

whose derivative is given by

$$g'(x) = 1 - \frac{f'(x)}{f'(x)} + \frac{f(x) f''(x)}{[f'(x)]^2}$$
$$= \frac{f(x) f''(x)}{[f'(x)]^2}$$

Since $f(s) = 0$, it follows that $g'(s) = 0$ as required which means Newton's method will converge quadratically *when it converges*.

Figure 5.5 shows a simple graphical representation of Newton's method. Notice how each x_i is the point where the tangent to $f(x)$ at x_{i-1} intersects the x-axis. Also, $\tan \theta_i = f'(x_i)$ whereby we can see from the trigonometric rules of a right-angled triangle that $x_i - x_{i+1} = f(x_0)/f'(x_0)$. Newton's method is widely used in computers as a basis for square root and reciprocal evaluation.

Example 6 Find the positive square root of a real number c by Newton's method.

To find $\sqrt{c}$, the nonlinear problem can be posed as

$$x^2 - c = 0.$$

The Newton iteration can then be expressed as

$$x_{n+1} = x_n - \frac{x_n^2 - c}{2x_n}$$
$$= \frac{x_n + c/x_n}{2}$$

Fig. 5.5 Applying Newton's method

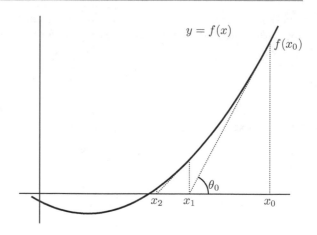

To demonstrate how the iteration would progress, let $c = 3$ and choose $x_0 = 1/2$. Then we get

$$x_1 = \frac{1/2 + 3/(1/2)}{2} = 3.25$$

$$x_2 = \frac{3.25 + 3/3.25}{2} = 2.0865385,$$

$$x_3 = 1.7621632$$

$$x_4 = 1.7320508$$

$$x_5 = 1.7320508$$

so with an initial guess perhaps not very close to the solution we have agreement in seven decimals after a fourth iteration. It's also worth noting that quadratic convergence can be observed by looking at the sequence of errors: 1.2321×10^0, 1.5179×10^0, 3.5449×10^{-1}, 3.0112×10^{-2}, 2.5729×10^{-4}, 1.9106×10^{-8}. Quadratic convergence can be observed by noting the exponent field doubles at each iteration.

Next we consider the implementation of Newton's method for solving $f(x) = 0$ in Python. The necessary inputs are the function f, its derivative, an initial guess and the required accuracy in the solution. In the implementation below, no limit is placed on the number of iterations. This would be needed for a robust implementation of Newton's (or any iterative) method. Our objective is not to create such software, but rather to get an idea of the basic ideas. Python and SciPy in particular have some robust equation-solvers built in. We shall discuss those briefly later.

Program   Python function for Newton's method

```
def newton(fcn, df, g, tol):
    """
    Solve the equation fcn(x)=0 to accuracy
    'tol'<1 using Newton's method. Use initial
    guess 'g'. 'fcn' and 'df' must be functions
    and the latter is the derivative of the
    former.

    """

    old = g + 1   # Ensure iteration starts.
    while abs(g - old) > tol:
        old = g
        g = old - fcn(old) / df(old)
    sol = g
```

Example 7   Solve the equation $e^x - 5x + 2 = 0$ using Newton's method.

We know that one solution lies in $[2, 3]$ and so take the initial guess 2.5. The function $e^x - 5x + 2$ and its derivative were defined as Python functions called **eq1** and **deq1** respectively after which the following Python command gave the result shown.

```
>>> s = newton(eq1, deq1, 2.5, 1e-10)
>>> print(s)
2.1937416678456176
```

To get an idea of the speed with which Newton's method found this solution, **g** was printed in each iteration in the Python command window. The successive iterates were:

```
2.2657507308869795
2.1990020960841123
2.1937726756131535
2.193741668931969
2.193741667845617
2.1937416678456176
```

Just six iterations were needed to obtain the result to high precision.

Fig. 5.6 Newton's method
does not always converge

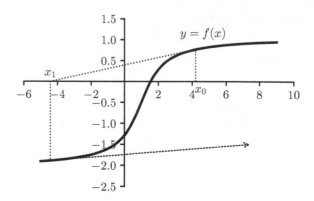

Upon looking at the iteration in general, it is clear that there are several things that could go wrong with the Newton iteration. Equation (5.7) means that if $f'(x_n)$ is small, then the correction made to the iterate x_n will be large, i.e. if the derivative of f is zero (or very small) near the solution, then Newton's method may not converge. Such a situation can arise if there are two solutions very close together, or, as in Fig. 5.6, when the gradient of f is small everywhere except very close to the solution.

Example 8 Consider the equation $\tan^{-1}(x - 1) = 0.5$ which has its unique solution $x = 1.5463$ to four decimals.

The function $f(x) = \tan^{-1}(x - 1) - 0.5$ is graphed in Fig. 5.6 along with the first two iterations of Newton's method with the rather poor starting point $x_0 = 4.2$ which yields $x_1 \approx -4.43$ and $x_2 \approx 53$. The oscillations get steadily wilder reaching `inf` and then `nan` in just ten more iterations using Python:

Iterate	Value
x_0	4.2
x_1	−4.43132
x_2	53.1732
x_3	−2810.47
x_4	1.63627e+07
x_5	−2.86693e+14
x_6	1.70205e+29
x_7	−3.10205e+58
x_8	1.99267e+117
x_9	−4.25185e+234
x_{10}	inf
x_{11}	nan

Of course, this particular equation can be solved very easily: rewrite it in the form $x = 1 + \tan(1/2) = 1.5463025$ to obtain the "solution". This rearrangement transforms the original problem from one of equation solving to one of function evaluation which is still a real computational problem as we shall see in the next chapter.

The following theorem explains the conditions under which we can expect Newton's method to converge. First, note that in Fig. 5.6, the function f is convex (concave up) to the left of the solution and concave (concave down) to the right– i.e it has an inflection point close to the solution causing issues. The global convergence theorem, Theorem 14, has hypotheses which eliminate the possibility of any points of inflection near the solution.

Theorem 14 *Let f be twice differentiable on an interval $[a, b]$ and satisfy the conditions*

(i) $f(a) f(b) < 0$

(ii) f' *has no zeros on* $[a, b]$

(iii) f'' *does not change sign in* $[a, b]$, *and*

(iv) $\left| \dfrac{f(a)}{f'(a)} \right|, \left| \dfrac{f(b)}{f'(b)} \right| < b - a$

Then $f(x) = 0$ has a unique solution $s \in (a, b)$ and Newton's iteration will converge to s from any starting point in $[a, b]$.

Proof The first condition establishes the existence of a solution since f must change sign in the interval. The second and third conditions force both f and f' to be strictly monotone. These guarantee the uniqueness of the solution and the absence of any inflection points. The final condition ensures that a Newton iteration from either endpoint a or b will generate a point in the interval (a, b). In combination these last two conditions ensure that all the iterates remain in the interval. The details of the proof are omitted. ∎

Our motivation for Newton's method was the desire to achieve quadratic convergence. In the next example provides the theoretical analysis of Newton's method for square roots, as in Example 6.

Example 9 Show that the iteration $x_{n+1} = \dfrac{x_n + c/x_n}{2}$ converges quadratically to $\sqrt{c}$ for any $x_0 > 0$.

Consider the sequence of errors defined by

$$e_n = x_n - \sqrt{c} \quad (n = 0, 1, 2, \ldots)$$

Note that $x_{n+1} = \left(x_n^2 + c \right) / 2x_n$ and so

$$x_{n+1} - \sqrt{c} = \frac{x_n^2 - 2x_n \sqrt{c} + c}{2x_n} = \frac{\left(x_n - \sqrt{c} \right)^2}{2x_n}$$

Similarly

$$x_{n+1} + \sqrt{c} = \frac{\left(x_n + \sqrt{c}\right)^2}{2x_n}$$

which implies that

$$\frac{x_{n+1} - \sqrt{c}}{x_{n+1} + \sqrt{c}} = \frac{\left(x_n - \sqrt{c}\right)^2}{\left(x_n + \sqrt{c}\right)^2} = \left(\frac{x_{n-1} - \sqrt{c}}{x_{n-1} + \sqrt{c}}\right)^4$$

$$= \cdots = \left(\frac{x_0 - \sqrt{c}}{x_0 + \sqrt{c}}\right)^{2^{n+1}}$$

Since $x_0 > 0$, it follows that $\left|\frac{x_0 - \sqrt{c}}{x_0 + \sqrt{c}}\right| < 1$ and hence $\frac{x_{n+1} - \sqrt{c}}{x_{n+1} + \sqrt{c}} \to 0$ as $n \to \infty$. There-
fore $x_n \to \sqrt{c}$ for every choice $x_0 > 0$.

Since even powers of real numbers are positive, it also follows from the analysis
above that $x_n > \sqrt{c}$ for every $n \geq 1$ so that

$$e_{n+1} = \frac{e_n^2}{2x_n} \leq \frac{e_n^2}{2\sqrt{c}}$$

which is to say the convergence is quadratic.

Exercises

1. The equation

$$3x^3 - 5x^2 - 4x + 4 = 0$$

   has a solution near $x = 0.7$. (See Exercise 1 in Sect. 5.2.) Carry out the first four
   iterations of Newton's method to obtain this solution.
2. Use Newton's method to obtain the solution of the equation $e^x - 3x - 1 = 0$ in
   $[1, 3]$ using the tolerance 10^{-12}.
3. Intervals containing the three solutions of $e^x - 100x^2 = 0$ were found in Exercise 3 of Sect. 5.2. Use Newton's method to locate the solutions with tolerance
   10^{-10}.
4. For the equation in Exercise 3, two solutions are close to $x = 0$. Try to find
   the critical value c such that if $x_0 > c$ then the solution near 0.1 is obtained,
   while if $x_0 < c$ the negative solution is located. Now try to justify your answers
   theoretically. (We are trying to find the *regions of attraction* for each of these
   solutions.)
5. Show that Newton's method for finding reciprocals by solving $1/x - c = 0$
   results in the iteration

$$x_{n+1} = x_n \left(2 - cx_n\right)$$

   Show that this iteration function satisfies $\left|g'(x)\right| < 1$ for $x \in (1/2c, 3/2c)$.

6. For the iteration in the previous exercise, prove that

$$x_{n+1} - \frac{1}{c} = -c \left(x_n - \frac{1}{c} \right)^2$$

Therefore $x_{n+1} < 1/c$. Show also that if $x_n < 1/c$ then $x_{n+1} > x_n$. It follows that, for $n \geq 1$, (x_n) is an increasing sequence which converges quadratically to $1/c$.

7. For Newton's iteration for finding $1/c$ with $c \in [1, 2)$ and $x_0 = 3/4$, show that six iterations will yield an error smaller than 2^{-65}.

8. Consider the coffee cooling scenario above. Use $T_E = 22^o C$, $T_0 = 90\,^\circ C$ and suppose your ideal temperature to drink the coffee is $\hat{T} = 80\,^\circ C$. Find the time that the coffee reaches that temperature by solving $\hat{T} - T(t) = 0$ if $k = 0.04$ using Newton's method and compare the convergence properties to your results with the bisection method. Experiment with different initial iterates.

5.5 The Secant Method

In the previous section, we saw that Newton's method is a powerful tool for solving equations, obtaining accurate solutions at a fast rate given the appropriate conditions. We also saw that some issues can arise. One challenge in the implementation of Newton's method is that derivatives are required. In many real-world applications, the derivative may not be available if, for example, the function itself is the result of some other computation. One important example of this considered later is in shooting methods for the solution of differential equations. The *secant method* is an approach to recover some of the power of Newton's method without using any derivative information.

Newton's method was described by the idea that the next iterate is the point at where the tangent line (at the current estimate of the solution) cuts the x-axis. The secant method uses the point at which the secant, or chord, line joining the two previous *two* iterates cross the x-axis. Figure 5.7 shows this.

To define the iteration, consider the equation of the secant line joining two points on the curve $y = f(x)$ at $x = x_0, x_1$. The slope is $\dfrac{f(x_1) - f(x_0)}{x_1 - x_0}$ and so equation of the secant line is

$$y - f(x_1) = \frac{f(x_1) - f(x_0)}{x_1 - x_0} (x - x_1). \tag{5.8}$$

Setting $y = 0$ in (5.8) and calling the solution x_2 gives the next iterate,

$$x_2 = x_1 - \frac{x_1 - x_0}{f(x_1) - f(x_0)} f(x_1).$$

Fig. 5.7 Applying the secant method

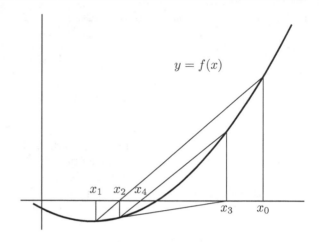

The general expression for the secant iterate is

$$x_{n+1} = x_n - \frac{x_n - x_{n-1}}{f(x_n) - f(x_{n-1})} f(x_n) \qquad (5.9)$$

Note this is exactly the Newton iteration with the true derivative replaced by the approximation

$$f'(x_n) \approx \frac{f(x_n) - f(x_{n-1})}{x_n - x_{n-1}},$$

which we know from the previous chapter on numerical differentiation may not be a good approximation! In this context, however, this simple approximation does the trick.

The secant method appears to have fast convergence but we can quantify the rate more precisely. There are convergence theorems for this method similar to those for Newton's method but they are beyond the scope of this book. The main conclusion is that when the secant method converges, it does so at a *superlinear* rate. Specifically, if e_n is the error $x_n - s$, then the sequence of errors satisfies

$$e_{n+1} \approx c e_n^\alpha$$

where $\alpha = \left(1 + \sqrt{5}\right)/2 \approx 1.6$. This can be interpreted as saying that the number of correct decimal places increases by about 60% with each iteration, compared to Newton's method in which the number of decimal places nearly doubles.

Example 10 The secant method can be applied to computing square roots.

If $f(x) = x^2 - c$ then the secant iteration becomes

$$x_{n+1} = x_n - \left(x_n^2 - c\right) \frac{x_n - x_{n-1}}{x_n^2 - x_{n-1}^2} = x_n - \frac{x_n^2 - c}{x_n + x_{n-1}}$$

$$= \frac{x_n^2 + x_n x_{n-1} - x_n^2 + c}{x_n + x_{n-1}} = \frac{x_n x_{n-1} + c}{x_n + x_{n-1}}$$

Recall, for Newton's iteration we had

$$x_{n+1} = \frac{x_n^2 + c}{2x_n} = \frac{x_n x_n + c}{x_n + x_n}$$

which is similar, especially when the iterates are close to the solution.

For $c = 3$, with the initial guesses $x_0 = 1/2$, $x_1 = 3/4$, the next few secant iterations are:

2.700000000000000
1.456521739130435
1.667887029288703
1.737709153632850
1.731944200430867
1.732050633712870
1.732050807574228
1.732050807568877.

Six iterations are accurate to 10 decimal places. So we still see fast convergence without having or using any derivative information.

Example 11 Consider the coffee model from Newton's law of cooling given by

$$T(t) = T_E + (T_0 - T_E)e^{-kt}.$$

Suppose you do not know the cooling rate k for your favorite mug, but you have temperature data for your coffee over time. Formulate a mathematical approach to estimating the model parameter k as a *nonlinear least squares* problem and use the secant method to approximate k. Use $T_E = 24.5$ and $T_0 = 93.5$ Use the temperature data given in the table below.

The idea is to try to approximate a value of k so that the coffee cooling model is "close" to the experimental data. This can be posed as

$$\min_k f(k) = \frac{1}{2} \sum_{i=1}^{20} (T_i(k) - \bar{T}_i)^2, \tag{5.10}$$

Times (min)	0	1	2	3	4	5	6	7
$T^{o}C$	93.50	76.9	65.9	54.6	45.3	40.3	37.9	32.95
Times (min)	8	9	10	11	12	13	14	15
$T^{o}C$	30.99	29.14	28.14	27.95	26.54	24.90	25.38	25.22
Times (min)	16	17	18	19	20			
$T^{o}C$	25.07	24.89	24.81	24.76	24.63			

where here $\bar{T}_i$ are the temperature measurements from above and T is a vector of model temperature values at the same times the experimental data was taken as a function of k. If we want to find the k value that minimizes $f(k)$ then we need to set $f'(k) = 0$, where this is the derivative *with respect to k*. So in the case of optimization, the nonlinear equation to be solved is actually in terms of the derivative of f, which is typically referred to as the objective function. Differentiating with respect to k gives the following nonlinear problem:

$$f'(k) = \sum_{i=1}^{20}(T_i(k) - \bar{T}_i - t_i(T_0 - T_E)e^{-kt_i}), \qquad (5.11)$$

where here we have applied the chain rule in expression Eq. (5.10) and the derivative of the coffee model expression with respect to k. Since this is a 1D problem, insight can easily be gained by looking at plots over a range of k values. This can help identify an initial iterate for example. Figure 5.8 shows the graph of the objective function and Fig. 5.9 shows the plot of Eq. (5.11). The problem appears to have a solution somewhere between 0.25 and 0.3.

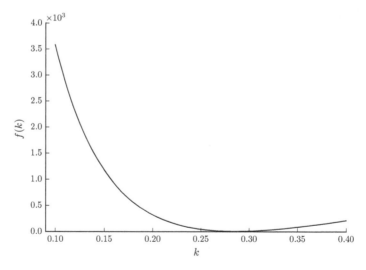

Fig. 5.8 Plot of the objective function over a range of k values to help see where the minimum occurs

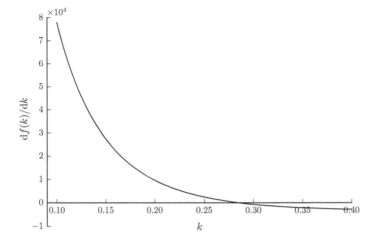

Fig. 5.9 Plot of the derivative for a range of k values. Where this graph is zero is the corresponding k value for the minimum of the objective function

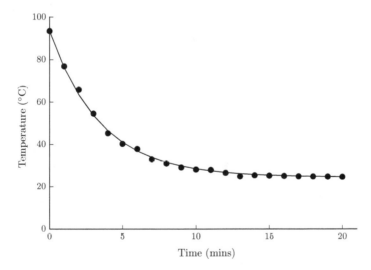

Fig. 5.10 Plot of the model output using the optimal value of k along with the experimental temperature values

The secant method requires two initial iterates. We will use 0.25 and 0.255 and a stopping criteria of 0.001. The secant algorithm gives the following iterates 0.2780, 0.2835, 0.2850, 0.2851 with a final objective function value of 199.1363 which has a derivative value of 0.0529. Figure 5.10 shows how the model using the optimal value of k agrees with the experimental temperature values.

Writing computer code for the secant method is left as an exercise, but should be straightforward using the Newton's method script to start from.

Exercises

1. The equation
$$3x^3 - 5x^2 - 4x + 4 = 0$$
   has a solution near $x = 0.7$. (See Exercise 1 in Sect. 5.2.) Carry out the first four iterations of the secant method to obtain this solution. Compare the results with those of Newton's method.
2. Write a program to solve an equation using the secant method. Test it by checking your answers to the previous exercises.
3. Use the secant method to obtain the solution of the equation $e^x - 5x + 2 = 0$ using the tolerance 10^{-12} and compare the results and convergence behavior to the previous examples.
4. Use the secant method to obtain the solution of the equation $e^x - 3x - 1 = 0$ in $[1, 3]$ using the tolerance 10^{-12}.
5. Intervals containing the three solutions of $e^x - 100x^2 = 0$ were found in Exercise 3 of Sect. 5.2. Use the secant method to locate the solutions with tolerance 10^{-10}. Compare the numbers of iterations used with those needed by Newton's method.
6. Show that the secant method for finding reciprocals by solving $1/x - c = 0$ results in a division-free iteration
$$x_{n+1} = x_n + x_{n-1} - cx_n x_{n-1}$$
   Carry out the first three iterations for finding $1/7$ using $x_0 = 0.1$, $x_1 = 0.2$.
7. For this exercise, we will use some ideas from calculus to model the curve of a hanging cable such as telephone wire and find its length. A cable hanging under its own weight assumes the shape of a catenary. In the special case that the ground is level and two poles are the same height, the lowest point on the curve will be at the midpoint of the distance between the two poles. Denote that point as $x = 0$ on the ground. The curve representing the shape of the cable for some parameters h_0 and λ is given by
$$y = h_0 + \lambda \cosh\left(\frac{x}{\lambda}\right). \tag{5.12}$$
   The length of the cable will depend on the shape of the catenary and the amount of sag in the middle. The amount of sag desired would be determined by engineers to allow for weather conditions in the area. The greater the sag, the less likely the cable will be blown down in a storm, for example.
   Suppose the following:
   • The two poles are 200 ft apart
   • The cable sags 15 ft in the middle
   • The height at which the cable is attached to the pole is 50 ft.

Find h_0, λ using the iterative methods presented in this chapter. To do this you need to determine an equation in one unknown (it will be nonlinear) and find the solution via the *bisection method, Newton's method, and the secant method.* Report on the iteration histories for each method. Finally, determine the length of the cable.

HINTS:

• The height at the midpoint is $y(0) = h_0 + \lambda$ and the height at an endpoint is $y(100) = h_0 + \lambda \cosh\left(\frac{100}{x}\right)$. How can you use this information and the sag condition to find one nonlinear equation in one unknown, $f(\lambda)$?

• Once you know λ how can you find h_0?

• The arc length of the resulting curve can be found using standard calculus techniques:

$$Length = 2 \int_0^{100} \sqrt{1 + \sinh^2\left(\frac{x}{\lambda}\right)} \, dx$$

Use elementary identities and properties of hyperbolic functions. (Ie. Look in your calc text.)

• You should get $\lambda = 335.7$ and $Length = 202.97$ ft.

5.6 Newton's Method in Higher Dimensions

The focus of this section is solving systems of nonlinear equations of the form

$$\mathbf{f}(\mathbf{x}) = \mathbf{0} \tag{5.13}$$

where $\mathbf{f}$ is a vector function of the vector variable $\mathbf{x}$. Later, we provide details on the specific case of two equations in two unknowns but we begin with the more general case of n equations in n unknowns.

Looking more closely at Eq. (5.13), we have

$$\mathbf{f}(\mathbf{x}) = \begin{bmatrix} f_1(x_1, x_2, \ldots, x_n) \\ f_2(x_1, x_2, \ldots, x_n) \\ \vdots \\ f_n(x_1, x_2, \ldots, x_n) \end{bmatrix} = \begin{bmatrix} 0 \\ 0 \\ \vdots \\ 0 \end{bmatrix} = \mathbf{0}$$

So the problem is now to find the n-dimensional vector $\mathbf{x} = [x_1, x_2, \ldots, x_n]^T$ that simultaneously satisfies the equations $f_i(\mathbf{x}) = 0$ for $i = 1, 2, \ldots n$. Deep insight into techniques requires some background in multivariable Calculus and Linear Algebra, but some of the key ideas are presented here. The thing to keep in mind is that when it comes to vector valued functions, some of the tools we used in solving $f(x) = 0$ for a scalar equation will no longer apply. For example, consider the Newton iterate defined by

$$x_+ = x_c - \frac{x_c}{f'(x_c)},$$

where here the subscripts $+$ and c refer to the next and current Newton iterate so as to not cause confusion with the subscripts in $x_i (i = 1, 2, \ldots, n)$ which are the components of $\mathbf{x}$.

What goes wrong when f is a function from R^n into R^n? We need to be careful about what we mean by a derivative first. In higher dimensions like this, the Taylor expansion of $f(\mathbf{x})$ actually leads to using a *Jacobian Matrix*, $\mathbf{J}$ defined in terms of the partial derivatives of the f_i,

$$J_{ij} = \frac{\partial f_i}{\partial x_j}.$$

So if the derivative is now a matrix, then dividing by $f'(x_c)$ isn't actually defined in that sense. Even more food for thought–what would go wrong with the secant approach?

However, Newton's method for such a system is still based on a first order Taylor expansion:

$$\mathbf{f}(\mathbf{x} + \mathbf{h}) \approx \mathbf{f}(\mathbf{x}) + J\mathbf{h} \tag{5.14}$$

where J is the Jacobian matrix of $\mathbf{f}$ evaluated at $\mathbf{x}$. As in the one-dimensional case, the Newton iteration is derived from setting the right hand side of (5.14) to $\mathbf{0}$. This leads to the iterative formula

$$\mathbf{x}_+ = \mathbf{x}_c - J^{-1}\mathbf{f}(\mathbf{x}_c). \tag{5.15}$$

In practice, the inverse of the Jacobian is not usually formed, rather Eq. (5.15) is re-arranged so that one computes the Newton step by solving the linear system

$$\mathbf{J}\mathbf{h} = -f(\mathbf{x}_c)$$

and then updating the current iterate with

$$\mathbf{x}_+ = \mathbf{x}_c + \mathbf{h}$$

Thus, the implementation of the iteration (5.15) requires the ability to solve general linear systems of equations, which you are already now familiar with. Also, fast quadratic convergence can still be maintained under analogous conditions.

Solving systems of nonlinear equations arises in the context of applied optimization problems. Across science and engineering disciplines, optimization problems abound. Decision makers want to minimize cost, minimize risk, maximize profits, or maximize customer satisfaction. Other times, within a design process–optimal parameters may be sought to fit a model to experimental data or to make a product give a desired output. In these instances, a mathematical representation of the stakeholder's 'goal' is called an objective function and such a problem can be posed as

$$\min_{x \in \Omega} f(\mathbf{x}) \tag{5.16}$$

but here f maps a vector of n decision variables to a real number. In Eq. (5.16), Ω defines a set of *constraints* on $\mathbf{x}$ that may be simple requirements like $x_i > 0$ for each i or they may be more complicated expressions that are themselves linear or nonlinear equations or inequalities. Some of those ideas may be familiar to you if you recall using LaGrange multipliers in a multivariable Calculus course. We will not go into much detail about that here, but to see how this is related to nonlinear equations, recall that a critical point in multivariable calculus means that all the partial derivatives of f are zero, i.e.

$$\nabla f \mathbf{x} = \mathbf{0},$$

where ∇f is the *gradient vector* with components defined by

$$\nabla f_i = \frac{\partial f}{\partial x_i}.$$

More care must be paid to whether or not the method is converging to a maximum or a minimum and most optimization algorithms have built-in mechanisms to ensure they are searching in an appropriate direction (i.e. the function values are decreasing as the optimization progresses).

Newton's method for optimization problems then requires solving the system of nonlinear equations given by Eq. (5.16) leading to

$$\mathbf{x}_+ = \mathbf{x}_c - H^{-1} \nabla f(\mathbf{x}_c). \tag{5.17}$$

where

$$H_{ij} = \frac{\partial^2 f}{\partial x_i \partial x_j}.$$

is the *Hessian matrix* of second partial derivatives. Also note that this matrix is naturally symmetric since mixed partial derivatives are equal.

Revisiting the coffee example from the previous section, there are certainly more accurate approaches than Newton's law of cooling. In reality, coffee cools quickly at the beginning (by a steaming mechanism) and after a while it cools mainly by conduction. To study the conduction mechanism, we could discard the hottest data when the steaming is most active, leaving say m data points. This means though, that we would need to estimate an initial temperature to use in the conduction-only approach. Now, the reformulated objective function also has the initial temperature as decision variable. So, we simultaneously seek k and T_0 for the following problem;

$$\min f(k, T_0) = \frac{1}{2} \sum_{i=1}^{m} T_i(k, T_0) - \hat{T}_i)^2. \tag{5.18}$$

Now the system of nonlinear equations determined by;

$$\nabla f(k, T_0) = \begin{bmatrix} \frac{\partial f}{\partial k} \\ \frac{\partial f}{\partial T_0} \end{bmatrix} = \begin{bmatrix} \sum_{i=1}^{m}(T_i(k, T_0) - \hat{T}_i)\left(\frac{dT_i}{dk}\right) \\ \sum_{i=1}^{m}(T_i(k, T_0) - \hat{T}_i)\left(\frac{dT_i}{dT_0}\right) \end{bmatrix}$$

and since $T(t) = T_E + (T_0 - T_E)e^{-kt}$, it follows that $\frac{dT_i}{dk} = -t_i(T_0 - T_E)e^{-kt}$ and $\frac{dT_i}{dT_0} = e^{-kt_i}$.

In the next section we provide the specific details about how to implement Newton's method for problems with two unknowns. Since the Jacobian matrix is only 2×2, it is straightforward to define an inverse. We should stress that the method is generalizable to any size problem of course, using many of the ideas presented in Chap. 4 on linear equations.

5.6.1 Newton's Method: Two Equations in Two Unknowns

We demonstrate the above ideas for $n = 2$, that is two equations in two unknowns. In this more manageable scenario, it is straightforward to get a formula to directly update the Newton iterate without having to solve a linear system at each iteration.

For two unknowns, the problem can be expressed as

$$f_1(x, y) = 0 \tag{5.19}$$
$$f_2(x, y) = 0$$

The first-order Taylor expansions of these functions gives

$$f_1(x + h, y + k) \approx f_1(x, y) + h f_{1x}(x, y) + k f_{1y}(x, y)$$
$$f_2(x + h, y + k) \approx f_2(x, y) + h f_{2x}(x, y) + k f_{2y}(x, y)$$

Setting the two right-hand sides (simultaneously) to zero leads to the solutions

$$h = \frac{-f_1 f_{2y} + f_2 f_{1y}}{f_{1x} f_{2y} - f_{1y} f_{2x}} \tag{5.20}$$

$$k = \frac{f_1 f_{2x} - f_2 f_{1x}}{f_{1x} f_{2y} - f_{1y} f_{2x}}$$

where all functions are to be evaluated at (x, y).

The Newton iteration for a pair of equations in two unknowns is therefore

$$x_{n+1} = x_n + h; \quad y_{n+1} = y_n + k \tag{5.21}$$

where (h, k) are given by (5.20) evaluated at (x_n, y_n).

Example 12 Find the coordinates of the intersections in the first quadrant of the ellipse $4x^2 + y^2 = 4$ and the curve $x^2 y^3 = 1$ illustrated below.

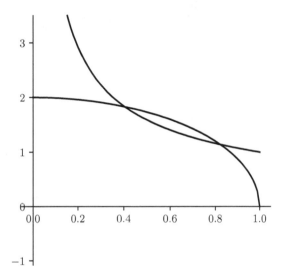

One possible technique for solving this pair of equations would be to eliminate x^2 between the two equations to obtain a fifth degree polynomial in y. This could in turn be solved using Newton's method for a single equation. Our purpose here is to illustrate the use of Newton's method for a pair of equations. The use of this substitution is left as an exercise to serve as a check on this process.

One of the solutions is close to $(0.4, 1.8)$ and we shall use this as a starting point for the iteration. The partial derivatives of the two functions $f_1(x, y) = 4x^2 + y^2 - 4$ and $f_2(x, y) = x^2 y^3 - 1$ are

$$f_{1x} = 8x, \quad f_{1y} = 2y$$
$$f_{2x} = 2xy^3, \quad f_{2y} = 3x^2 y^2$$

At the point $\mathbf{x}_0 = (0.4, 1.8)$, we have

$$f_1 = 4(0.4)^2 + 1.8^2 - 4 = -0.12$$
$$f_2 = (0.4)^2 (1.8)^3 - 1 = -0.06688$$

and

$$f_{1x} = 3.2, \quad f_{1y} = 3.6$$
$$f_{2x} = 4.665\,6, \quad f_{2y} = 1.555\,2$$

Hence $f_{1x}f_{2y} - f_{1y}f_{2x} = (3.2)(1.555\,2) - (3.6)(4.665\,6) = -11.819\,52$. Now applying (5.20) we get

$$h = \frac{-(-0.12)(1.555\,2) + (-0.0\,668\,8)(3.6)}{-11.819\,52} = 4.580\,896\,7 \times 10^{-3}$$

$$k = \frac{(-0.12)(4.665\,6) - (-0.0\,668\,8)(3.2)}{-11.819\,52} = 2.926\,142\,5 \times 10^{-2}$$

so that

$$x_1 = 0.4 + 4.580\,896\,7 \times 10^{-3} = 0.404\,580\,9$$
$$y_1 = 1.8 + 2.926\,142\,5 \times 10^{-2} = 1.829\,261\,4$$

Subsequent iterations generate the points (0.4041495644206274, 1.829386038547648), (0.4041494570206883, 1.8293859258121536), (0.4041494570206442, 1.8293859258121765).

A typical iteration of this method for this pair of equations can be implemented in Python using:

```
f1 = 4 * x0**2 + y0**2 - 4
f2 = x0**2 * y0**3 - 1
f1x = 8 * x0
f1y = 2 * y0
f2x = 2 * x0*y0**3
f2y = 3 * x0**2 * y0**2
D = f1x * f2y - f1y * f2x
h = (f2 * f1y - f1 * f2y) / D
k = (f1 * f2x - f2 * f1x) / D
x0 = x0 + h
y0 = y0 + k
```

The other solution is close to $(0.8, 1.2)$. Using this starting point, we get the iterates (0.8229861111111112, 1.1387037037037038), (0.8216842595962869, 1.1398895102304403), (0.8216816252038416, 1.1398935157519317), (0.8216816251900977, 1.1398935157723458), (0.8216816251900978, 1.1398935157723458), and (0.8216816251900978, 1.1398935157723458) again.

A very small number of iterations has provided both solutions to very high accuracy.

Newton's method is fairly easy to implement for the case of two equations in two unknowns. We first need to define functions for the equations and the partial derivatives. For the equations in Example 2, these can be given by:

```python
import numpy as np

def eq2(v):
    """
    The function's input 'v' and the output are
    both 2-entry vectors.

    """

    x, y = v
    f = np.empty(2)
    f[0] = 4 * x**2 + y**2 - 4
    f[1] = x**2 * y**3 - 1
    return f
```

and

```python
def Deq2(v):
    """
    Compute Jacobian for the function 'eq2'.

    """

    x, y = v
    J = np.empty((2,2))
    J[0,0] = 8 * x
    J[0,1] = 1 * y;
    J[1,0] = 1 * x * y**3
    J[1,1] = 3 * x**2 * y**2
    return J
```

The newton2 function will need both the function and its partial derivatives as well as a starting vector and a tolerance. The following code can be used.

Program   Python function for Newton's method for two equations

```
def newton2(fcn, Jac, g, tol):
    """
    Solve two equations given by fcn(x)=0 to
    accuracy 'tol' < 1 using Newton's method. 'g'
    is an initial guess. 'fcn' and 'Jac' must be
    functions where the latter is the partial
    derivatives of the former.

    """

    old = np.zeros_like(g)
    old[0] = g[0] + 1
    while max(abs(g - old)) > tol:
        old = g
        f = fcn(old)
        f1 = f[0]
        f2 = f[1]
        J = Jac(old)
        f1x = J[0,0]
        f1y = J[0,1]
        f2x = J[1,0]
        f2y = J[1,1]
        D = f1x * f2y - f1y * f2x
        h = (f2 * f1y - f1 * f2y) / D
        k = (f1 * f2x - f2 * f1x) / D
        g = old + np.array((h, k))
    return g
```

Then the following command can be used to generate the second of the solutions in Example 12:

```
>>> s = newton2(eq2, Deq2, [.8, 1.2], 1e-8)
>>> print(s)
[ 0.82168162   1.13989352]
```

Exercises

1. Eliminate x^2 between the equations $4x^2 + y^2 = 4$ and $x^2 y^3 = 1$ to get a polynomial equation in y for the intersection points of these curves.(See Example 12.)
2. Use appropriate starting points to find the y coordinates of the intersection points of the curves in Exercise 1 using Newton's method for a single equation. Verify that these solutions yield the same intersection points as were found in Example 12.

Fig. 5.11 Curves from
Exercise 3

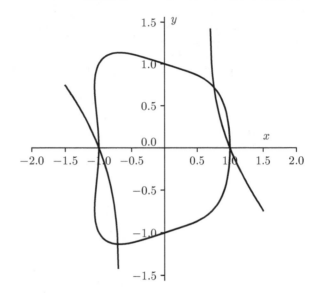

3. In Fig. 5.11 are graphs of the curves defined by the equations $x^4 + xy^2 + y^4 = 1$
   and $x^2 + xy - y^2/4 = 1$. There are two obvious intersection points at $(\pm 1, 0)$.
   Use an appropriate starting point and perform the first two iterations of Newton's
   method for the other intersection in the first quadrant.
4. For the curves in the previous exercise, there are intersection points close to
   $(0.8, 0.8)$ and $(-0.75, -1)$. Use Newton's method to find these intersection
   points to high accuracy.
5. Consider the hanging cable model from Exercise 7 in Sect. 5.5. It is not realistic
   that the ground is level. We can modify the above model to take into account hills
   and valleys as in the figure below (Fig. 5.12).
   Let A be the left endpoint and B be the right endpoint and suppose the following:
   • B is 50 ft higher than A
   • A and B are 1 mile apart (5280 ft) and poles are located every 220 ft. Note that
   each arc of the cable will be a catenary.
   • There is a hill 150 ft high 1/4 mile from A and a valley whose floor is 80 ft
   below B 1/3 mile from B (as in the picture)
   • Note that each arc of the cable will be a catenary. We require that each arc sag
   10 ft below the straight line at the vertical midpoints of the heights.
   • We require that no point on the cable be less than 25 ft above the ground.
   • The ground heights relative to A are given in the table below.
   Find the length of the entire cable. This will include finding the governing non-
   linear equation for the unknown parameters (you will have a system of two equa-
   tions and two unknowns) and generalizing Newton's method to two dimensions.
   Finally, plot the curve of the cable and the ground profile.

Distance from A	0	220	440	660	880	110	1320	1540	1760
Height above A	0	20	49	83	115	140	150	135	112
Distance from A	1980	2200	2420	2640	2860	3080	3300	3520	3740
Height Above A	88	63	39	16	−3	−18	−27	−30	−28
Distance from A	3960	4180	4400	4620	4840	5060	5280		
Height above A	−25	−21	−15	−6	−7	26	50		

Fig. 5.12 Depiction of uneven ground for cable design. Note the vertical scale is exaggerated to highlight the terrain

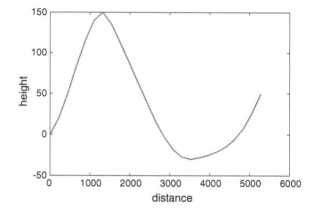

HINTS:

• First, consider one arc of the cable with endpoints of different heights. Suppose the two endpoints are $x = 0$ and $x = L$. The cable will still hang under the shape of a catenary and the equation for the shape of the curve is given by

$$y = h + \lambda \cosh \frac{x + c}{\lambda} \tag{5.22}$$

where we need to determine the appropriate h, λ, and c. We can eliminate h as an unknown to obtain our system of two equations in two unknowns, λ and c.

• The heights at the endpoints are $y(0) = h + \cosh \frac{c}{\lambda}$ and $y(L) = h + \cosh \frac{L+c}{\lambda}$. The difference in these heights is the same as the difference between the ground levels at the poles (which is a known value)! So one equation is

$$\lambda \cosh \frac{L + c}{\lambda} - \lambda \cosh \frac{c}{\lambda} = y(L) - y(0). \tag{5.23}$$

The second equation can be determined from the sag condition. We can now use Newton'e method for two equations and two unknowns to find λ and c.

• To get your code running, try this for the simple case when the ground is level for $L = 220$. In this case $y(0) - y(L) = 0$ and symmetry yields $c = -110$. Using Newton's method for scalar equations would give $\lambda = 606.66$ so a good initial guess for your 2-D code (to see that it is working) would be $c = -100$, $\lambda = 600$. Try as an initial guess for the uneven terrain $c = 2(y(L) - y(0)) - 110$ and $\lambda = 600$.

- You need use Newton's method on each interval to find the corresponding c, λ values of each catenary. You can then compute the arc length on each interval and sum them for the total length of the cable.
- You need to determine h. This must be done relative to some h_0 so that the cable is continuous and h_0 must be determined so that the cable is always kept 25 ft above the ground.

6. When on a trip to the carnival with your friends, your best friend gets on the Ferris wheel, accidentally forgetting their candy apple. Being the great friend that you are, you decide to throw it to them. You've only got one apple and one chance to make the perfect shot to deliver your friend their delicious treat. Luckily, you have taken Calculus, and know that you can calculate the perfect moment to make your throw by using the parametric representation of the apple trajectory and the equations of motion of the Ferris wheel. For the apple the trajectory is given by

$$x = f(t) = 22 - 8.03(t - t_0),$$
$$y = g(t) = 1 + 11.7(t - t_0) - 4.9(t - t_0)^2,$$

while the Ferris wheel moves according to

$$x = h(t) = 15 \sin\left(\frac{\pi t}{10}\right)$$
$$y = l(t) = 16 - 15 \cos\left(\frac{\pi t}{10}\right).$$

Here t_0 is the time you launch the apple. As you start calculating, you realize that you have the worst throwing arm of anyone you know. It may take you several iterations to get this right! Here is your methodology!

(a) Given the apple trajectory, find the speed and angle in relation to t at which the apple is thrown when $t = t_0$.

(b) Now that you know the angle and speed of the candy apple, you need to find out where your friend is. So, using the Ferris wheel equation, find the location of your friend at $t = 0$ (Hint: They should be at the bottom of the Ferris wheel).

(c) Find the number of rotations per minute of the ferris wheel.

(d) Set up a mathematical formulation for this scenario that represents what you are trying to do (throw the apple and have your friend catch it). You will have two unknowns: the initial time to throw the candy apple, and the time when your friend will receive the apple.

(e) This is a system of two nonlinear equations with two unknowns. Use Newton's method to solve this problem. It may help to play around and try a few different values of the initial launch time and apple-catch time to understand how close and far you are.

7. For this problem, you will consider the optimal shape and construction of a trash
   dumpster.

   (a) First, locate a dumpster. Carefully study the dimensions and describe all
       aspects of the dumpster. Second, determine the volume of the dumpster.
       Include a sketch of the dumpster. Digital pictures can accompany, but a
       sketch with labeled components is required. Your job is to find the dimen-
       sions of a container with the same volume that minimizes construction costs.
       For example, something like in Fig. 5.13
   (b) While maintaining the same general shape and method of construction,
       determine the dimensions such a container with the same volume should
       have in order to minimize the cost of construction. Review from Calculus
       how you find the minimum of a function of several variables.
       Use the following assumptions in your analysis:
       • Again, the volume of the dumpster must be the same.
       • The angle of the top of the dumpster must be preserved. Relate the height
       of the back panel to the front panel, then use the volume constraint to sim-
       plify this further
       • The sides, back, and front are to made of 12-gauge (0.1046 inch thick)
       steel sheets, which cost $0.87 per square foot (including cuts or bends)
       • The base is made from a 10-gauge (0.1345 inch thick) steel sheet which
       costs $1.79 per square foot
       • Lids cost approximately $85.00 each regardless of dimensions
       • Welding costs approximately $0.37 per foot of material with labor included

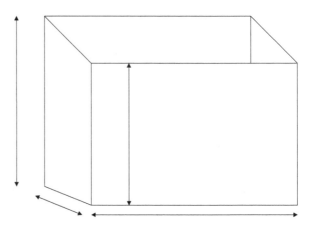

Fig. 5.13 Example
dumpster and dimensions to
be measured

(c) Describe how any of your assumptions or simplifications may effect your design. For example, are there any fixed costs you did not include in the design? Would they affect the validity of your results. If you were hired as a consultant on this investigation, what would your conclusion be? Would you recommend altering the design of the dumpster? If so, describe the savings that would result. It is important be as specific as possible when explaining your procedures, mathematical model, experiments, measurements, and computations.

8. Consider solving the refined coffee model problem in which the period of "steaming" is ignored. To do this, disregard all the data that is hotter than $65\,^\circ$C from data given in Example 11.

   (a) Write a subroutine to evaluate the gradient function.
   (b) Implement Newton's method for this new problem. You will need to either determine the Jacobian of ∇f *or* consider using finite difference Jacobians (i.e. Hessian matrices)
   (c) Experiment with different initial iterate pairs of (k, T_0) values. How does your choice impact the convergence results. Provide some illustrative plots to demonstrate your solution.
   (d) Is there a significant improvement over the original formulation? What are other sources of error that may be impacting your fit?

5.7 Conclusions and Connections: Iterative Solution of Nonlinear Equations

What have you learned in Chap. 5? And where does it lead?

This section offered many choices in approaching scalar nonlinear equations, $f(x) = 0$. A significant difference between these problems and the methods presented for linear equations is that they are inherently iterative. Care *must* be taken at multiple stages of the solution process in order to obtain a reasonable solution. Something as simple as having a good starting point for the iteration can be the difference between convergence or divergence. In practice, when f represents a real-world application or is modeling a physical phenomenon then intuition or inside perspective should be used to choose the starting point instead of using a random number. Moreover, methods like the bisection method can be used to generate a starting point for faster methods, such as the secant or Newton's methods.

In additional to the starting point, knowing when to *stop* the iteration is critical as well. In this chapter, we did not go into great detail about stopping criteria but they are also choices a user must make in implementing any of the methods here. In practice, stopping an algorithm based on $|f(x)|$ (which we hope is close to zero) or a relative reduction (i.e. $|f(x)|/f(x_0)|$) or a combination of those two are all commonplace, as well as prescribing a fixed number of iterations before exiting the algorithm. It is also important to incorporate stopping criteria that represent failure since we cannot necessarily know in advance whether a chosen iteration and starting point will lead to convergence at all. Leaving your computer in a futile infinite loop of meaningless calculations is something to be avoided!

Knowing information about your problem can also help guide choices in terms of which algorithm to choose in general. In the absence of derivatives, the secant method still maintains fast convergence. We should note that using a finite difference approximate derivative (for example, something from Chap. 3) within Newton's method is also a possibility. Knowing the rate of convergence of a method is a way of testing your code. By solving a problem you know the answer to, if you look at successive errors, you will see the exponent field changing accordingly.

Finally, we presented the setting for systems of nonlinear equations. Both the model for a hanging cable exercise and the coffee cooling model parameter estimation example have been extended from one unknown to two. The details for implementing Newton's method for two equations and two unknowns are provided (in addition to the scalar algorithms) to so that you have the solvers all ready to go and tackle the applied problems in this chapter.

Several issues pertaining to solution of single equations with one unknown have been discussed. The introduction to Newton's method in two variables gives just a glimpse of a major field. Systems of several (or even many) equations in many unknowns present a major challenge–and one that arises frequently.

Some numerical methods themselves are built from a basic structure but with weights or coefficients that are not known, so-called undetermined coefficients. Frequently these must be found by solving systems of equations–sometimes linear, often nonlinear. Mathematical models for physical or other situations often depend on model parameters which need to be estimated from data. Again this typically results in systems of equations. We saw a little of that in the previous chapter with linear least squares approximation leading to systems of linear equations. Often the appropriate model does not depend linearly on the parameters and so the problem becomes one of systems of nonlinear equations.

Many real world problems are posed in a way that seeks a "best" answer by some definition of best. Typically this results in a mathematical formulation that requires optimization. Optimization problems can be of (essentially) infinite dimension where the objective is to find the function that (say) minimizes energy consumption. Many times the resulting optimization problem becomes one of minimizing a function of many variables, perhaps subject to some constraints on the variables. In Calculus you saw that such problems can be recast as nonlinear equations by setting all partial derivatives to zero. Optimization is much bigger than this, but what we see is that solution of nonlinear equations is a far-reaching topic with many aspects that have not be explored here.

There are still many opportunities for you to contribute in the future!

5.8   Python Functions for Equation Solving

For many of the tasks we shall discuss including the solution of equations, Python has functions using efficient algorithms in the third-party packages SciPy and NumPy The optimize module in the scipy package provides several functions for equation solving.

fsolve   In the case of solving nonlinear equations of a single variable, the basic Python (SciPy) function is fsolve.

This function takes two required arguments, the function f and the starting point x_0. It then finds the solution of $f(x) = 0$ nearest to x_0. Optional arguments are possible extra arguments to the function f, the partial derivatives of f, the tolerance, as well as several additional parameters to configure fsolve if desired.

The fsolve function uses the Powell hybrid method (not discussed in this book). (Newton's method cannot be used in this general context because of its need for the derivative.)

Example 13   We use SciPy's fsolve function to solve $e^x - 2x - 1 = 0$ with tolerance 10^{-8}. (Recall that this function has already been used. It was defined as eq1.)

The command

```
>>> s = fsolve(eq1, 1, xtol=1e-8)
>>> print(s)
```

gives the output

```
[ 1.25643121]
```

fsolve can also solve multi-variate equations, i.e. where the function takes a vector as input. Another alternative to fsolve in scipy.optimize is the function root which lets you choose among a number of different solution methods.

For more details on these and other SciPy functions, consult the SciPy documentation or type help(scipy.optimize) at the Python prompt after importing scipy.optimize.

The numpy package also provides useful functions for equation solving.

roots   For the special case of finding roots of polynomial equations, we can use roots.

This function will find the roots of the polynomial equation $p(x) = 0$ where p is represented in Python by a NumPy array of its coefficients, beginning with the highest degree term. The polynomial $x^2 + 2x - 3$ is therefore represented by the array $[1, 2, -3]$. Any zero coefficients must be included.

Example 14 Solve the polynomial equation $y^5 - 4y^3 + 4 = 0$ (see Exercise 1 of Sect. 5.6)

The polynomial is represented by the coefficient vector $[1, 0, -4, 0, 0, 4]$. The Python command

```
>>> roots([1,0,-4,0,0,4])
```

gives the output

```
[-2.10454850+0.j
  1.82938593+0.j
  1.13989352+0.j
 -0.43236547+0.85117985j
 -0.43236547-0.85117985j]
```

Notice that the real and complex roots are all listed. Note too that the two positive real roots coincide with our Newton's method solution to the original curve intersection problem in Example 2

For systems of nonlinear equations there is no simple equivalent of fzero. However there is the function fmins which implements an algorithm for the minimization of a function of several variables. The connection between these two problems is that solving a system of equations $f_1(\mathbf{x}) = f_2(\mathbf{x}) = \cdots = f_n(\mathbf{x}) = 0$ is equivalent to locating the minimum of the function

$$F(\mathbf{x}) = \sum_{k=1}^{n} |f_k(\mathbf{x})|^2$$

The details of the algorithms used are topics for more advanced courses.

Interpolation

6

6.1 Introduction

What is interpolation and why do we need it? The short answer to "What is interpolation?" is that we seek to find a function of a particular form that goes through some specified data points. Our focus will be entirely on single variable interpolation. Much of what we will be doing is based on polynomial interpolation in which we seek a polynomial of specified degree that agrees with given data.

So, why do we need interpolation? Many of the standard elementary and special functions of mathematics cannot be evaluated exactly–even if we assume a computer with infinite arithmetic precision. Even where we have good algorithms to approximate these functions they may be very expensive to compute. Thus we seek simpler functions that agree with the data and which can be used as surrogates for the underlying function. Of course in most practical situations we don't even know what that underlying function is!

Interpolation is a powerful tool used in industry, often used to replace complex, computationally expensive functions with friendlier, cheaper ones. In engineering design, interpolation is often used to create functions (usually low degree polynomials) so that experimental data can be incorporated into a continuous model. This idea is *similar* to least-squares approximations (as in Chap. 4 or with regards to the coffee model) but instead of looking for a polynomial that might "fit" data closely, the interpolating polynomial is required to equal, or agree with, the data at certain locations. When data is subject to error (experimental error, for example) it is common to use approximation methods which force the approximating function to pass close to all the data points without necessarily going through them.

One special interpolation technique, called spline interpolation, is the basis of much computer aided design. In particular, Bezier splines were developed originally for the French automobile industry in the early 1960s for this specific purpose. Bezier splines are a generalization of the cubic splines to general planar curves. They are

© Springer International Publishing AG, part of Springer Nature 2018   189
P. R. Turner et al., *Applied Scientific Computing*, Texts in Computer Science,
https://doi.org/10.1007/978-3-319-89575-8_6

not necessarily interpolation splines but make use of what are called "control points" which are used to "pull" a curve in a particular direction. Many computer graphics programs use similar techniques for generating smooth curves through particular points.

Spline interpolation was an important tool in creating the early digitally animated movies such as Toy Story. The ability to have visually smooth curves that do not deteriorate under scaling for example is critical to creating good images. Another image processing application of the basic idea is the removal of unwanted pieces in a photograph. We remove the telephone wire, for example, and need to fill the void with appropriate shape and color information. One approach to this is interpolation to ensure a smooth transition across such an artificial boundary. Digital photography relies on interpolation to complete the image from the basic red, green or blue receptors.

First we provide some background and justification for choosing polynomials as the basic tools for approximating functions. Note that any continuous function can be approximated to any required accuracy by a polynomial (we provide the theorem below) and surprisingly polynomials are the only functions, which can, theoretically at least, be evaluated exactly.

The first of these reasons is based on a famous theorem of Weierstrass.

Theorem 15 (Weierstrass) *Let f be a continuous function on the interval $[a, b]$. Given any $h > 0$, there exists a polynomial $p_{N(h)}$ of degree $N(h)$ such that*

$$\left| f(x) - p_{N(h)}(x) \right| < h$$

for all $x \in [a, b]$. Therefore there exists a sequence of polynomials such that $\|f - p_n\|_\infty \to 0$ as $n \to \infty$.

The second reason for choosing polynomials is less obvious. To gain some insight we briefly describe *Horner's rule* for efficient evaluation of a polynomial.

Suppose we'd like to evaluate

$$p(x) = a_n x^n + a_{n-1} x^{n-1} + \cdots + a_1 x + a_0$$

Horner's rule states that

$$p(x) = \left\{ \cdots \left[(a_n x + a_{n-1}) x + a_{n-2} \right] x + \cdots + a_1 \right\} x + a_0$$

This is easily implemented in Python using the following code:

```python
def horner(a, x):
    """

    Evaluate the polynomial using Horner's rule:
    p(x) = a(0) + a(1)x + ... + a(n-1)x^(n-1)

    """
```

```
n = len(a)
p = a[-1]
for k in range(2, n+1):
    p = p * x + a[-k]
return p
```

Note that indexing arrays with a negative index as above (a[-k]) in Python indexes the array from the back end. That is, the last entry in an array can be accessed as a[-1], the next-to-last as a[-2] etc.

For example the polynomial $-x^2 + 2x + 4$ could be evaluated at $x = 3$ by

```
>>> horner([4, 2, -1], 3)
```

which returns the expected value 1. NumPy's function polyval is essentially similar – except that it has the coefficient vector in the reverse order.

Considering this as a piece of hand calculation, we see a significant saving of effort. In the code above each of the $n - 1$ steps entails a multiplication and an addition, so that for a degree $n - 1$ polynomial, a total of $n - 1$ additions and $n - 1$ multiplications are needed while direct computation requires more. Evaluation of $a_k x^k$ requires k multiplications, and therefore the complete operation would need

$$(n - 1) + (n - 2) + \cdots + 2 + 1 = \frac{n(n-1)}{2}$$

multiplications and $n - 1$ additions. Horner's rule is also numerically more stable, though this is not obvious without resorting to highly pathological examples.

Example 1 Use Horner's rule to evaluate

$$p(x) = x^5 + 2x^4 + 3x^3 + 4x^2 + 5x + 6$$

for $x = 0.4321$.

We get

$$\begin{aligned}
p(x) &= (\{[(x + 2)x + 3]x + 4\}x + 5)x + 6 \\
&= (((((0.4321) + 2)(0.4321) + 3)(0.4321) + 4)(0.4321) + 5)(0.4321) + 6 \\
&= 9.234159
\end{aligned}$$

To six decimals the sequence of values generated is:
1, 2.4321, 4.050910, 5.750398, 7.484747, 9.234159

We see that polynomials offer us both ease of evaluation and arbitrary accuracy of approximation, making them a natural starting point for interpolation.

Exercises

1. Use Horner's rule to evaluate $\sum_{k=0}^{10} kx^{10-k}$ for $x = 0.7$.
2. Repeat the previous exercise using the function horner and using NumPy's polyval command.
3. Modify the horner for NumPy array inputs. Use it to evaluate the polynomial in Exercise 1 for $x = -1 : 0.1 : 2$. Use the resulting data to plot a graph of this polynomial.

6.2   Lagrange Interpolation

Polynomial interpolation is the basic notion that we build a polynomial which agrees with the data from the function f of interest (or simply data). The *Lagrange interpolation polynomial* has the property that it takes the same values as f at a finite set of distinct points.

Actually, this idea isn't entirely new to you! The first $N + 1$ terms of the Taylor expansion of f about a point x_0 form a polynomial, the *Taylor polynomial*, of degree N,

$$p(x) = f(x_0) + (x - x_0) f'(x_0) + \cdots + \frac{(x - x_0)^N}{N!} f^{(N)}(x_0)$$

which satisfies the *interpolation conditions*

$$p^{(k)}(x_0) = f^{(k)}(x_0) \qquad (k = 0, 1, \ldots, N)$$

Taylor's theorem also provides the error in using $p(x)$ to approximate $f(x)$ is given by

$$f(x) - p(x) = \frac{(x - x_0)^{N+1}}{(N + 1)!} f^{(N+1)}(\xi)$$

for some ξ lying between x and x_0.

Note that in many data-fitting applications, we are only given function values and the corresponding derivative values are not available; secondly, this approach will usually provide good approximations only near the base point x_0. So using Taylor polynomials exclusively just isn't enough.

The Taylor polynomial does however illustrate the general polynomial interpolation approach: first find a formula for the polynomial (of minimum degree) which satisfies the required interpolation conditions, and then find an expression for the error of the resulting approximation. In reality of course, that error can't be computed explicitly since, if it were, then we could evaluate the function exactly in the first place.

To develop the theory of the Lagrange interpolation polynomial, consider the values of a function f at $N + 1$ distinct points, called *nodes*, $x_0, x_1, \ldots, x_N$. The problem now is to find the minimum degree polynomial p such that

$$p(x_k) = f(x_k) \qquad (k = 0, 1, \ldots, N) \tag{6.1}$$

Suppose that p has degree m. Then we can write

$$p(x) = a_m x^m + a_{m-1} x^{m-1} + \cdots + a_1 x + a_0 \tag{6.2}$$

where the coefficients $a_0, a_1, \ldots, a_m$ need to be determined. Substituting (6.2) into (6.1) yields the following system of linear equations for these unknown coefficients:

$$
\begin{aligned}
a_0 + a_1 x_0 + \cdots + a_{m-1} x_0^{m-1} + a_m x_0^m &= f(x_0) \\
a_0 + a_1 x_1 + \cdots + a_{m-1} x_1^{m-1} + a_m x_1^m &= f(x_1) \\
&\vdots \\
a_0 + a_1 x_N + \cdots + a_{m-1} x_N^{m-1} + a_m x_N^m &= f(x_N)
\end{aligned}
\tag{6.3}
$$

In general, if $m < N$, then the linear system will have no solution. It will have infinitely many solutions if $m > N$ but, provided the matrix A with elements $a_{ij} = x_{i-1}^{j-1}$ is nonsingular (which it is), a unique solution exists if $m = N$. We thus expect our interpolation polynomial to have degree N (or less if it turns out that $a_N = 0$). Although Eq. (6.3) provides us with a theoretical way to find coefficients, a more practical method to finding the polynomial p satisfying (6.1) is needed.

To see how this is achieved, we consider a more specific goal: find polynomials l_j $(j = 0, 1, \ldots, N)$ of degree at most N such that

$$l_j(x_k) = \delta_{jk} = \begin{cases} 1 & \text{if } j = k \\ 0 & \text{if } j \neq k \end{cases} \tag{6.4}$$

and then the polynomial p given by

$$p(x) = \sum_{j=0}^{N} f(x_j) l_j(x) \tag{6.5}$$

will have degree at most N and will satisfy the interpolation conditions (6.1).

Before obtaining these polynomials explicitly, you can see that p defined by (6.5) satisfies the desired interpolation conditions;

$$
\begin{aligned}
p(x_k) &= \sum_{j=0}^{N} f(x_j) l_j(x_k) = \sum_{j=0}^{N} f(x_j) \delta_{jk} \\
&= f(x_k)
\end{aligned}
$$

by (6.4).

Next, the requirement that $l_j(x_k) = 0$ whenever $j \neq k$ means that $l_j(x)$ must have factors $(x - x_k)$ for each such k. There are N such factors so that

$$l_j(x) = c \prod_{k \neq j} (x - x_k) = c(x - x_0) \cdots (x - x_{j-1}) \cdot (x - x_{j+1}) \cdots (x - x_N)$$

has degree N and satisfies $l_j(x_k) = 0$ for $k \neq j$. It remains to choose the constant c so that $l_j(x_j) = 1$. This requirement implies that

$$c = \frac{1}{\displaystyle\prod_{k \neq j} (x_j - x_k)}$$

so that

$$l_j(x) = \prod_{k \neq j} \frac{(x - x_k)}{(x_j - x_k)} = \frac{(x - x_0) \cdots (x - x_{j-1}) \cdot (x - x_{j+1}) \cdots (x - x_N)}{(x_j - x_0) \cdots (x_j - x_{j-1}) \cdot (x_j - x_{j+1}) \cdots (x_j - x_N)}$$

$$\tag{6.6}$$

These polynomials are called the *Lagrange basis polynomials* and the polynomial p given by (6.5) is the *Lagrange interpolation polynomial*.

To this end, the existence of this polynomial has been established by finding it, although the linear equations (6.3) indicated this as well. These same considerations also show the uniqueness of the interpolation polynomial of degree at most N satisfying (6.1), which relies on the fact that the matrix A mentioned above is nonsingular. The proof of this, by showing that the *Vandermonde determinant*

$$V = \begin{vmatrix} 1 & x_0 & x_0^2 & \cdots & x_0^N \\ 1 & x_1 & x_1^2 & \cdots & x_1^N \\ & \cdots & & & \cdots \\ 1 & x_N & x_N^2 & \cdots & x_N^N \end{vmatrix}$$

is nonzero for distinct nodes $x_0, x_1, \ldots, x_N$, is left as an exercise. An alternative proof of uniqueness can be obtained from the fundamental theorem of algebra since we cannot find two *distinct* polynomials p and q of degree N which have the same values at $N + 1$ distinct nodes. (This would imply that the difference $p - q$ would have $N + 1$ distinct roots.)

Example 2 Find the Lagrange interpolation polynomial for the following data and use it to estimate $f(1.4)$.

x	1	2	3
$f(x)$	1.0000	1.4142	1.7321

With $x_0 = 1$, $x_1 = 2$, and $x_2 = 3$, we obtain the quadratic interpolation polynomial given by

$$p(x) = f(x_0) \frac{(x - x_1)(x - x_2)}{(x_0 - x_1)(x_0 - x_2)} + f(x_1) \frac{(x - x_0)(x - x_2)}{(x_1 - x_0)(x_1 - x_2)}$$

$$+ f(x_2) \frac{(x - x_0)(x - x_1)}{(x_2 - x_0)(x_2 - x_1)}$$

$$= 1.0000 \frac{(x - 2)(x - 3)}{(1 - 2)(1 - 3)} + 1.4142 \frac{(x - 1)(x - 3)}{(2 - 1)(2 - 3)}$$

$$+ 1.7321 \frac{(x - 1)(x - 2)}{(3 - 1)(3 - 2)}$$

Evaluating $p(x)$ at $x = 1.4$,

$$p(1.4) = 1.0000 \frac{(1.4 - 2)(1.4 - 3)}{(1 - 2)(1 - 3)} + 1.4142 \frac{(1.4 - 1)(1.4 - 3)}{(2 - 1)(2 - 3)} + 1.7321 \frac{(1.4 - 1)(1.4 - 2)}{(3 - 1)(3 - 2)}$$

which gives gives $f(1.4) \approx 1.1772$ to 4 decimal places. Note that for this data, $f(x) = \sqrt{x}$ and so the true value is $\sqrt{1.4} = 1.1832$ giving an absolute error of 6.0160×10^{-3}.

With any approximation, an expression for the error guides practical implementation. The proof of the error formula is accomplished with the repeated use of Rolle's theorem which states that between any two zeros of a differentiable function there must be a point at which the derivative is zero.

Theorem 16 (Error in Lagrange interpolation) *Suppose the function f is $N + 1$ times continuously differentiable on the interval $[a, b]$ which contains the distinct nodes $x_0, x_1, \ldots, x_N$. Let p be the Lagrange interpolation polynomial given by Eqs. (6.5) and (6.6). Then, for any $x \in [a, b]$,*

$$f(x) - p(x) = \frac{(x - x_0)(x - x_1) \cdots (x - x_N)}{(N + 1)!} f^{(N+1)}(\xi) \tag{6.7}$$

for some $\xi \in [a, b]$.

Proof Firstly, for simplicity of notation, let

$$L(x) = \prod_{k=0}^{N} (x - x_k) = (x - x_0)(x - x_1) \cdots (x - x_N)$$

Now, for $x \in [a, b]$, but not one of the nodes, consider the function E defined by

$$E(t) = f(t) - p(t) - cL(t)$$

where the *constant* c is chosen so that $E(x) = 0$. It now follows that $E(t)$ vanishes at the $N + 2$ distinct points $x_0, x_1, \ldots, x_N$, and x. By Rolle's theorem, between each successive pair of these there is a point at which E' vanishes. Repeating this argument for $E', E'', \ldots, E^{(N)}$ implies that there is a point $\xi \in [a, b]$ such that

$$E^{(N+1)}(\xi) = 0$$

From the definition of E, this yields

$$f^{(N+1)}(\xi) - p^{(N+1)}(\xi) - cL^{(N+1)}(\xi) = 0$$

But, p is a polynomial of degree at most N, and so $p^{(N+1)} \equiv 0$, and $L^{(N+1)} \equiv (N+1)!$ It follows that

$$c = f^{(N+1)}(\xi) / (N+1)!$$

which, combined with the fact that $E(x) = 0$, completes the proof. ∎

Note that Theorem 16 remains valid for x outside $[a, b]$ with a minor adjustment to the range of possible values of ξ. In such a case, this theorem would cease being about *interpolation*. The process of *extrapolation* in which we attempt to obtain values of a function outside the spread of the data is numerically much less satisfactory since the error increases rapidly as x moves away from the interval $[a, b]$.

To see this, it is sufficient to study a graph of the function $L(x)$. In Fig. 6.1 we have a graph of this function for the nodes $1, 1.5, 2, 2.5, 3$ so that

$$L(x) = (x - 1)(x - 1.5)(x - 2)(x - 2.5)(x - 3)$$

We see that the magnitude of this function remains small for $x \in [1, 3]$, but that it grows rapidly outside this interval. If $f^{(N+1)}$ varies only slowly then this is an

Fig. 6.1 Illustration of how the Lagrange error grows further from the nodes

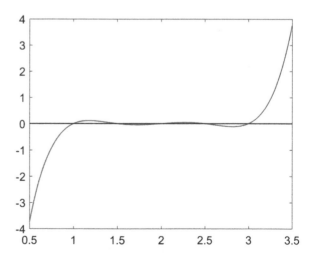

accurate reflection of the behavior of the error term for Lagrange interpolation using these same nodes.

Intuitively, one would expect the accuracy to improve as the number of nodes increases. Theorem 16 bears this out to some extent since the $(N + 1)!$ is likely to dominate the numerator in (6.7) *provided that* the higher-order derivatives of f remain bounded. Unfortunately, in practice, this may not be the case. Moreover, the inclusion of additional data points in the Lagrange interpolation formula is not completely straightforward. See the following example.

Example 3 Repeat the computation of Example 2 with the additional data point $f(1.2) = 1.0954$.

Since we now have a new data point, the previous calculations don't help and $p(x)$ must be reconstructed. Taking the nodes in numerical order, the new polynomial is

$$p(x) = 1\frac{(x - 1.2)(x - 2)(x - 3)}{(1 - 1.2)(-1 - 2)(1 - 3)} + 1.0954\frac{(x - 1)(x - 2)(x - 3)}{(1.2 - 1)(1.2 - 2)(1.2 - 3)}$$
$$+ 1.4142\frac{(x - 1)(x - 1.2)(x - 3)}{(2 - 1)(2 - 1.2)(2 - 3)} + 1.7321\frac{(x - 1)(x - 1.2)(x - 2)}{(3 - 1)(3 - 1.2)(3 - 2)}.$$

and so $p(1.4) = 1.1837$ to four decimal places giving an absolute error of 4.8404×10^{-4} (since the true function was $f(x) = \sqrt{x}$ and $\sqrt{1.4} = 1.1832$ to four decimal places).

From (6.7), the bound for this error as follows.

$$\sqrt{1.4} - p(1.4) = \frac{(0.4)(0.2)(-0.6)(-1.6)}{4!}f^{(4)}(\xi)$$

for some ξ between 1.2 and 2. In this case, $f^{(4)}(x) = (-15/16)x^{-7/2}$ and for $x \in [1., 2]$, it follows that that $|f^{(4)}(x)| \leq 1/2$. It follows that the error is bounded by $\frac{0.0768}{24}(0.5) = 1.6000 \times 10^{-3}$.

The true error is actually smaller than the predicted error since we computed the error bound using a maximum bound on $|f^{(4)}(x)|$.

There are many different ways of representing this interpolation polynomial, some of which we propose in the next section. The principal motivation is to find convenient forms which can be implemented efficiently and are easily programmed. The Lagrange form does not fulfill either of these criteria. However, The most important aspects of the Lagrange interpolation polynomial are that it provides a relatively straightforward way to prove the existence and uniqueness of the polynomial. Moreover, it allows for a simpler derivation of the error formula. Since the polynomial itself is unique, all the alternatives are just different ways of writing down the same thing–so there is no need to repeat the error analysis for the various forms.

As well as being able to choose the representation of the interpolation polynomial, in some circumstances, one may be in a position to choose the nodes themselves. Consideration of the remainder term (6.7) shows that it could be advantageous to choose the interpolation points so that $|L(x)|$ is kept as small as possible over the whole interval of interest. This motivation leads to the choice of the Chebyshev interpolation points. For details of Chebyshev interpolation, the reader is referred to a more advanced numerical analysis text. Several such are listed among the references and further reading.

Exercises

1. Find the Lagrange interpolation polynomial for the data

$$
\begin{array}{c|ccc}
x & 1 & 3 & 4 \\
\hline
f(x) & 3 & 1 & 1
\end{array}
$$

   Use this to estimate $f(2.5)$.
2. Repeat Exercise 1 with the additional data $f(0) = 4$, $f(2) = 2$.
3. Show that the Vandermonde determinant

$$
V = \begin{vmatrix}
1 & x_0 & x_0^2 & \cdots & x_0^N \\
1 & x_1 & x_1^2 & \cdots & x_1^N \\
& \cdots & & & \cdots \\
1 & x_N & x_N^2 & \cdots & x_N^N
\end{vmatrix}
$$

   does not vanish for distinct nodes $x_0, x_1, \ldots, x_N$. (**Hint:** Show that $(x_j - x_k)$ is a factor of V whenever $j \neq k$.)
4. Consider the following table of values of the sine function:

$$
\begin{array}{c|ccccccccc}
x & 0.0 & 0.1 & 0.2 & 0.3 & 0.4 & 0.5 & 0.6 & 0.7 & 0.8 \\
\hline
\sin x & 0.0000 & 0.0998 & 0.1987 & 0.2955 & 0.3894 & 0.4794 & 0.5646 & 0.6442 & 0.7174
\end{array}
$$

   Write down the Lagrange interpolation polynomial using the nodes $0.0, 0.1$, and 0.2, and then using 0.3 as well. Estimate the value of $\sin 0.26$ using each of these polynomials.
5. Obtain error bounds for the approximations in Exercise 4.
6. Show that for any $x \in (0.0, 0.8)$, the Lagrange interpolation quadratic using the three nearest nodes from the table in Exercise 4 results in an error less than 6.5×10^{-5}.
7. Show that if $x \in (0.1, 0.7)$ then the Lagrange interpolation polynomial using the four closest points in the table in Exercise 4 results in an error less than 2.5×10^{-6}.
8. Repeat Exercise 5 for the function $\ln(1 - x)$ for the same nodes. How many points will be needed to ensure that the Lagrange interpolation polynomial will introduce no new errors to four decimal places?

6.3 Difference Representations

In this section, the focus turns to alternative representations of interpolation polynomials which may be more convenient and efficient to program, and to use in practice. All the representations considered here can be derived from the the use of *divided differences*. We begin with Newton's divided difference representation and then briefly explore some *finite difference* interpolation formulas that are special cases of the divided difference formula for the situation where all the nodes are equally spaced.

6.3.1 Divided Difference Interpolation

Before discussing their use in interpolation, we must define divided differences.

Definition 17 *The* zero-th order divided difference of f at x_k *is defined by*

$$f[x_k] = f(x_k)$$

First-order divided differences *at two points* x_j, x_k *are then defined by*

$$f[x_j, x_k] = \frac{f[x_k] - f[x_j]}{x_k - x_j} = f[x_k, x_j] \tag{6.8}$$

Higher-order divided differences are defined recursively by

$$f[x_k, x_{k+1}, \ldots, x_{k+m}] = \frac{f[x_{k+1}, \ldots, x_{k+m}] - f[x_k, \ldots, x_{k+m-1}]}{x_{k+m} - x_k} \tag{6.9}$$

Note that (6.8) implies a connection between first divided differences and first derivatives. Recall that the secant iteration in Chap. 5, used the approximation

$$f'(x_n) \approx \frac{f(x_n) - f(x_{n-1})}{x_n - x_{n-1}}$$

which is to say

$$f'(x_n) \approx f[x_{n-1}, x_n]$$

Further, by the mean value theorem, it follows that

$$f[x_{n-1}, x_n] = \frac{f(x_n) - f(x_{n-1})}{x_n - x_{n-1}} = f'(\xi)$$

for some point ξ between x_{n-1} and x_n. Soon we will show that there is a more general connection between divided differences and derivatives of the same order.

Example 4 Compute the differences $f[x_0, x_1]$, $f[x_0, x_2]$, $f[x_1, x_2]$, $f[x_0, x_1, x_2]$, and $f[x_1, x_0, x_2]$ for the data

$$
\begin{array}{c|ccc}
k & 0 & 1 & 2 \\
x_k & 2 & 3 & 5 \\
f(x_k) & 4 & 2 & 1
\end{array}
$$

Applying (6.8),

$$
f[x_0, x_1] = \frac{f[x_1] - f[x_0]}{x_1 - x_0} = \frac{2 - 4}{3 - 2} = -2
$$

$$
f[x_0, x_2] = \frac{f[x_2] - f[x_0]}{x_2 - x_0} = \frac{1 - 4}{5 - 2} = -1
$$

$$
f[x_1, x_2] = \frac{f[x_2] - f[x_1]}{x_2 - x_1} = \frac{1 - 2}{5 - 3} = -\frac{1}{2}
$$

Now using (6.9)

$$
f[x_0, x_1, x_2] = \frac{f[x_1, x_2] - f[x_0, x_1]}{x_2 - x_0} = \frac{-1/2 - (-2)}{3} = \frac{1}{2}
$$

$$
f[x_1, x_0, x_2] = \frac{f[x_0, x_2] - f[x_1, x_0]}{x_2 - x_1} = \frac{-1 - (-2)}{2} = \frac{1}{2}
$$

Note that in Example 4, $f[x_0, x_1, x_2] = f[x_1, x_0, x_2]$. This *independence of order* is a general property of divided differences. Direct proofs of this property tend to be conceptually simpler but algebraically more intricate than the one provided later. The interested reader can consult an advanced numerical analysis text for alternative proofs. One approach will be described here and is based on the uniqueness of the interpolating polynomial. First, the basics of divided difference interpolation need to be established.

From the definition of first-order divided differences (6.8), we see that

$$
f(x) = f[x_0] + (x - x_0)f[x_0, x] \tag{6.10}
$$

Substituting (6.9) into (6.10) with $k = 0$ and using increasing values of m, we then get

$$
\begin{aligned}
f(x) &= f[x_0] + (x - x_0)f[x_0, x] \\
&= f[x_0] + (x - x_0)f[x_0, x_1] + (x - x_0)(x - x_1)f[x_0, x_1, x] \\
&= \cdots \\
&= f[x_0] + (x - x_0)f[x_0, x_1] + (x - x_0)(x - x_1)f[x_0, x_1, x_2] \\
&\quad + \cdots + (x - x_0)\cdots(x - x_{N-1})f[x_0, x_1, \ldots, x_N] \\
&\quad + (x - x_0)\cdots(x - x_N)f[x_0, x_1, \ldots, x_N, x]
\end{aligned} \tag{6.11}
$$

If we now let p be the *polynomial* consisting of all but the last term of this expression:

$$p(x) = f[x_0] + (x - x_0) f[x_0, x_1] + (x - x_0)(x - x_1) f[x_0, x_1, x_2]$$
$$+ \cdots + (x - x_0) \cdots (x - x_{N-1}) f[x_0, x_1, \ldots, x_N] \qquad (6.12)$$

then p is a polynomial of degree at most N, and if $0 \le k \le N$, we obtain

$$p(x_k) = f[x_0] + (x_k - x_0) f[x_0, x_1] + (x_k - x_0)(x_k - x_1) f[x_0, x_1, x_2]$$
$$+ \cdots + (x_k - x_0) \cdots (x_k - x_{k-1}) f[x_0, x_1, \ldots, x_k] \qquad (6.13)$$

since all higher-degree terms have the factor $(x_k - x_k)$.

This enables us to prove the following theorem.

Theorem 18 *The polynomial given by (6.12) satisfies the interpolation conditions*

$$p(x_k) = f(x_k) \qquad (k = 0, 1, \ldots, N)$$

Proof We shall proceed by induction. First equation (6.13), with $k = 0$ reduces to just

$$p(x_0) = f[x_0] = f(x_0)$$

as desired. Next suppose the result holds for $k = 0, 1, \ldots, m$ for some $m < N$. Using (6.11) with $N = m$ and $x = x_{m+1}$ we get

$$f(x_{m+1}) = f[x_0] + (x_{m+1} - x_0) f[x_0, x_1] + (x_{m+1} - x_0)(x_{m+1} - x_1) f[x_0, x_1, x_2]$$
$$+ \cdots + (x_{m+1} - x_0) \cdots (x_{m+1} - x_{m-1}) f[x_0, x_1, \ldots, x_m]$$
$$+ (x_{m+1} - x_0) \cdots (x_{m+1} - x_m) f[x_0, x_1, \ldots, x_m, x_{m+1}]$$

which, using (6.13), is $p(x_{m+1})$. This completes the induction step and, therefore, the proof. ∎

Recall that by the uniqueness of the Lagrange interpolation polynomial, this formula (6.12) is just a rearrangement of the Lagrange polynomial. This special form is known as *Newton's divided difference interpolation polynomial*. It is a particularly useful form of the polynomial as it allows the data points to be introduced one at a time without any of the waste of effort this entails for the Lagrange formula.

Another implication of (6.12) is that the divided differences $f[x_0, x_1, \ldots, x_k]$ is the leading (degree k) coefficient of the interpolation polynomial which agrees with f at the nodes $x_0, x_1, \ldots, x_k$. By the uniqueness of this polynomial, this coefficient must be independent of the order of the nodes. This observation means that divided differences depend only on the set of nodes, not on their order.

The uniqueness of the interpolation polynomial also provides an important connection between divided differences and derivatives. We've seen that first divided differences are also approximations to first derivatives, or, using the mean value theorem, *are* first derivatives at some "mean value" point.

Subtracting Eq. (6.12) from (6.11), we obtain

$$f(x) - p(x) = (x - x_0) \cdots (x - x_N) f[x_0, x_1, \ldots, x_N, x]$$

but from Theorem 16, and, specifically, Eq. (6.7), we already know that

$$f(x) - p(x) = \frac{(x - x_0)(x - x_1) \cdots (x - x_N)}{(N + 1)!} f^{(N+1)}(\xi)$$

It follows that

$$f[x_0, x_1, \ldots, x_N, x] = \frac{f^{(N+1)}(\xi)}{(N + 1)!}$$

where the "mean value point" ξ lies somewhere in the interval spanned by $x_0, x_1, \ldots, x_N$, and x.

Example 5 Use Newton's divided difference formula to estimate $\sqrt{0.3}$ and $\sqrt{1.1}$ from the following data

x	0.2	0.6	1.2	1.5
$\sqrt{x}$	0.4472	0.7746	1.0954	1.2247

Taking the data in the order given results in the following table of divided differences.

k	x_k	$f[x_k]$	$f[x_k, x_{k+1}]$	$f[x_k, x_{k+1}, x_{k+2}]$	$f[x_k, x_{k+1}, x_{k+2}, x_{k+3}]$
0	0.2	0.4472	0.8185	−0.2837	0.1296
1	0.6	0.7746	0.5347	−0.1153	
2	1.2	1.0954	0.4310		
3	1.5	1.2247			

Here, for example, the entry $f[x_1, x_2] = 0.5347$ results from $(1.0954 - 0.7746)/(1.2 - 0.6)$. So for $x = 0.3$, from the table of differences we obtain the successive approximations, here shown rounded to four decimal places

$$f[x_0] + (x - x_0) f[x_0, x_1] = 0.4472 + (0.3 - 0.2)0.8185$$
$$= 0.5291$$

and then,

$$0.5291 + (x - x_0)(x - x_1) f[x_0, x_1, x_2]$$
$$= 0.5291 + (0.3 - 0.2)(0.3 - 0.6)(-0.2837)$$
$$= 0.5376$$

and, finally,

$$0.5376 + (0.3 - 0.2)(0.3 - 0.6)(0.3 - 1.2)(0.1296) = 0.5411$$

One immediately apparent aspect of this process is the ease of introducing new data points into the calculation, and the possibility of proceeding in an iterative manner until successive approximations to the required value agree to within some specified accuracy. This provides a much more practical way of performing polynomial interpolation which avoids the need for finding an error bound – often a difficult task in practice.

For $\sqrt{1.1}$, we change the order of the data to use data closest to the point of interest first. So we use the data in the order 1.2, 1.5, 0.6 and then 0.2 The resulting table of differences is then

k	x_k	$f[x_k]$	$f[x_k, x_{k+1}]$	$f[x_k, x_{k+1}, x_{k+2}]$	$f[x_k, x_{k+1}, x_{k+2}, x_{k+3}]$
0	1.2	1.0954	0.4310	−0.1153	0.1296
1	1.5	1.2247	0.5002	−0.2448	
2	0.6	0.7746	0.8185		
3	0.2	0.4472			

The successive approximations thus obtained are

$$1.0523, \quad 1.0477, \quad 1.0503$$

The true value is $\sqrt{1.1} = 1.0488$ to four decimal places. Here we see that the introduction of faraway data at the final stage harms the accuracy of the approximation. This helps emphasize the merits of using nearby data first, and perhaps restricting the number of data point used.

Note that data closer to the point of interest is in general more relevant and using the original order would mean we base our approximation on furthest data first. It's also more prone to accumulated error since the closest data only enters through higher order divided differences. The following program for divided difference interpolation reorders the data for each point at which we wish to estimate the function.

Program Python function for divided difference interpolation using nearest nodes first.

```python
import numpy as np

def ddiff1(xdat, ydat, x):
    """
    Function for computing Newton's divided
    difference formula using nearest data points
    in 'xdat', 'ydat' to elements of 'x' first.

    """

    N = x.size
    M = xdat.size
    D = np.zeros((M, M))
    y = np.zeros(N)
    for k in range(N):
        # Sort contents of input data arrays
        xtst = x[k]
        ind = np.argsort(np.abs(xtst-xdat))
        xsort = xdat[ind]
        D[:, 0] = ydat[ind]
        # Begin divided differences
        for j in range(M):
            for i in range(M-j-1):
                D[i,j+1] = ((D[i+1, j] - D[i,j])
                    / (xsort[i+j+1] - xsort[i]))
        # End divided differences
        # Compute interpolation
        xdiff = xtst - xsort
        prod = 1  # Holds the product of xdiff
        for i in range(M):
            y[k] = y[k] + prod * D[0,i]
            prod = prod * xdiff[i]
    return y, D
```

For the data of Example 5, the following Python commands yield the results shown.

```python
>>> X = np.array([0.2, 0.6, 1.2, 1.5])
>>> Y = np.sqrt(X)
>>> x = np.array([0.3, 1.1])
>>> y = ddiff1(X, Y, x)
>>> print(y)
[ 0.54106893  1.05032546]
```

Fig. 6.2 Divided difference interpolation

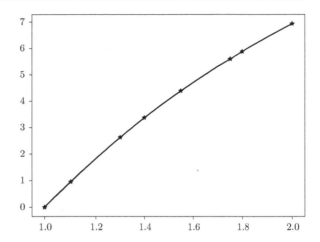

as we obtained earlier.

If your problem has many data points it may be desirable to limit the degree of the interpolation polynomial. This can be incorporated into the code by restricting the variable M to be bounded by some maximum degree.

To evaluate Newton's formula at many points (in order to graph the function, for example) it may be economical to avoid re-sorting the data for each argument. If the order of the data is not to be changed, then it is typically the case that *all* data points are used throughout. However if the number of data points is also large, the resulting graph may exhibit the natural tendency of a polynomial to "wiggle".

Example 6 Graph the divided difference polynomial for the data

$$x \quad 1.0000 \ 1.1000 \ 1.3000 \ 1.4000 \ 1.5500 \ 1.7500 \ 1.8000 \ 2.0000$$
$$y \quad 0.0000 \ 0.9530 \ 2.6240 \ 3.3650 \ 4.3830 \ 5.5960 \ 5.8780 \ 6.9310$$

over the interval $[1, 2]$.

With data vectors X,Y and x=1:0.01:2, we can use the Python commands

```
>>> y = ddiff1 (X, Y, x)
>>> plt.plot (X, Y, 'k*', x, y, 'k')
>>> plt.show ()
```

These generate the graph shown in Fig. 6.2.

Fig. 6.3 Interpolation of the
Runge function

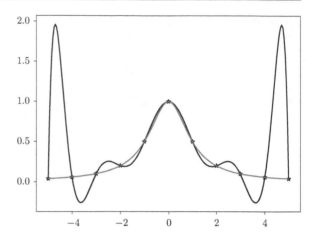

As mentioned above, for graphing purposes it may be desirable to avoid re-sorting the data. The code used previously can be modified for this purpose. (See the Exercises.) The computation for the next example uses such a function.

Example 7 Graph the interpolation polynomial for data from the function

$$y = \frac{1}{1 + x^2}$$

using nodes at the integers in $[-5, 5]$.

The Python commands

```
>>> X = np.arange(-5, 5+1)
>>> Y = runge(X)
>>> x = np.linspace(-5, 5, int(1/0.05 + 1.5))
>>> y = ddiff1(X, Y, x)
>>> plt.plot(X, Y, 'k*', x, y, 'k')
>>> plt.plot(x, runge(x))
>>> plt.show()
```

where the function runge implements the given function, producing the graph shown in blue in Fig. 6.3.

The interpolation polynomial and the "true" function become quite distinct towards the ends of the range, although they have reasonable resemblance to each other in the middle of the interval but as $|x|$ grows the polynomial behavior of the interpolation formula cannot be controlled.

This is a famous example from Runge. If the number of (equally-spaced) nodes is increased the oscillations of the polynomial become steadily wilder. Runge actually showed that, for this example, the interpolation polynomials of increasing degree *diverge* for $|x| > 3.5$.

Fig. 6.4 Local cubic interpolation of the Runge function

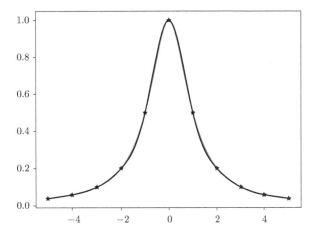

If the degree of the interpolating polynomial is restricted so that the nearest four nodes (to the current point) are used to generate a local cubic interpolating polynomial, then the resulting curve fits the original function much better. The commands

```
>>> plt.plot(X, Y, 'k*', x, runge(x))
>>> y1 = interp1d(X, Y, kind='cubic')
>>> plt.plot(x, y1(x),'k')
>>> plt.show()
```

produce the graph shown in Fig. 6.4. At a glance, the two appear to agree well.

The function interp1d is an interpolation function imported from scipy.interpolate, which will be revisited later.

In the last example, the data points were equally-spaced and so the divided difference formula can be rewritten in special forms. There are efficient ways of implementing divided difference interpolation to reduce the computational complexity. Among these, perhaps the most widely used is the algorithm of Aitken, which computes the values of the interpolation polynomials directly without the need for *explicit* computation of the divided differences. The details of Aitken's algorithm are left to subsequent, more advanced courses. We turn our attention to the special case of finite difference interpolation for data at equally-spaced nodes.

In introducing this approach to finite difference interpolation, we assumed that x_0 was chosen so that $x \in (x_0, x_1)$. If x lies much closer to x_1 than to x_0, then the order of the nodes is not exactly as we desired. In such a situation, it would be preferable to choose x_0 so that $x \in (x_{-1}, x_0)$ and then use the data points in the order $x_0, x_{-1}, x_1, x_{-2}, \ldots$ The examples above have demonstrated some practical difficulties with polynomial interpolation. Locally restricting the degree of the polynomial used can result in better approximation to a smooth function – but such an approximation does not reflect the smoothness of the function being approximated. The next section on spline interpolation presents a way to addresses this difficulty.

Exercises

1. Compute the divided differences $f[x_0, x_1]$, $f[x_0, x_2]$, and $f[x_0, x_1, x_2]$ for the data

k	0	1	2
x_k	0	2	5
$f(x_k)$	1	3	7

2. Write down the quadratic polynomial p which has the values $p(0) = 1$, $p(2) = 3$, $p(5) = 7$. (**Hint:** use Exercise 1.)
3. Use the divided difference interpolation formula to estimate $f(0.97)$ from the data

$$f(0.7) = 5.132, \quad f(0.9) = 5.343, \quad f(1.0) = 5.646$$
$$f(1.1) = 5.746, \quad f(1.3) = 5.939$$

4. Use divided difference interpolation to obtain values of $\cos 0.17$, $\cos 0.45$ and $\cos 0.63$ from the following data

x	0.0	0.1	0.2	0.3	0.4
$\cos x$	1.0000	0.9950	0.9801	0.9553	0.9211

x	0.5	0.6	0.7	0.8
$\cos x$	0.8776	0.8253	0.7648	0.6967

5. Write a script for divided difference interpolation which uses all the data, in the same order, at all points. Use this to plot the divided difference interpolation polynomial for the data in Exercise 4. Also plot the error function for this interpolation polynomial over the interval $[0, 1]$. (Note the behavior in the *extrapolation* region.)
6. Modify your code from Exercise 5 so that you use data closest to the point of interest first, and can restrict the degree of the local polynomial being used. Test your code by computing the interpolating function and the error function over the interval $[0, 1.2]$.

6.4 Splines

The previous section demonstrated ways to build polynomials to agree with a function or data at specified locations. It was also shown that using a single polynomial is not always a good choice in approximating values between those points. Using divided difference or finite difference formulas alleviated some of the pitfalls by using only those data points that are close to the current point of interest. However, it is sometimes advantageous to approximate the function (or interpolate data) using *piecewise polynomials* on different intervals.

The basic idea is probably not new to you, especially at the simplest level of piecewise linear functions. For example, suppose you look at the temperature forecast for the day and see it is going to be 15 °C at 10:00AM and 19 °C at 12:00 PM, then you may consider predicting the temperature at 11:00AM by thinking about connecting those 'data points' with a line. You can determine the line between $(10, 15)$ and $(12, 19)$ to get $y = 2x - 5$ (where y is the temperature and x is the time) so at $x = 11$ it would predict 17 °C. Here, there was an assumption made that the temperature would behave linearly between times. Also, you are not using any information about what the temperature was at say 6:00AM. Instead, you are just trusting the line between those two data points.

For data with more complicated behavior (or curvature), higher degree polynomial on each interval may be a better choice. The problem with the piecewise polynomials generated by, say, divided difference interpolation using cubic pieces (which would mean using the four nearest nodes to each point) is that the transition from one piece to the next is typically not smooth. This situation is illustrated in Fig. 6.5. The nodes used were 0, 1, 1.25, 1.5, 1.75, 2, 3. The data are plotted along with the divided difference piecewise cubic using the nearest four nodes. There is an abrupt change (indeed a discontinuity) at the point at which the node 0 is dropped from the set being used. For these nodes, this happens as soon as $|x - 1.75| < |x|$, which occurs at $7/8$.

The so-called *spline* interpolation can eliminate such loss of smoothness at the transitions.

Definition 19 *Let* $a = x_0 < x_1 < \cdots < x_n = b$. *A function* $s : [a, b] \to \Re$ *is a* spline *or* spline function *of degree* m *with* knots *(or* nodes, *or* interpolation points*)* $x_0, x_1, \ldots, x_n$ *if*

Fig. 6.5 Interpolation by a piecewise cubic polynomial

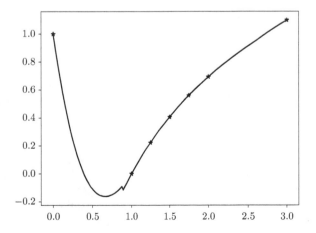

1. *s is a piecewise polynomial such that, on each subinterval* $[x_k, x_{k+1}]$, *s has degree at most m, and*
2. *s is m − 1 times differentiable everywhere.*

Thus, a spline function is defined by a different polynomial formula on each of the "knot intervals" $[x_k, x_{k+1}]$. So a spline function is differentiable at all points between the knots with the additional condition is that its first $m - 1$ derivatives are also continuous at each knot.

Example 8 Show that the following function is a spline of degree 1 (i.e. a linear spline)

$$s(x) = \begin{cases} x + 1 & x \in [-1, 0] \\ 2x + 1 & x \in [0, 1] \\ x + 2 & x \in [1, 3] \end{cases}$$

All that is needed is to show that s is continuous at the internal knots 0, 1. Now

$$\text{As } x \to 0^- \ s(x) \to 1$$
$$\text{As } x \to 0^+ \ s(x) \to 1$$

and so s continuous at $x = 0$. Similarly

$$\text{As } x \to 1^- \ s(x) \to 3$$
$$\text{As } x \to 1^+ \ s(x) \to 3$$

establishing the desired continuity at $x = 1$.

Example 9 Is the following function a spline?

$$s(x) = \begin{cases} x^3 & 0 \le x \le 1 \\ 2x^2 - 2x + 1 & 1 \le x \le 2 \\ 3x^2 - 6x + 5 & 2 \le x \le 4 \end{cases}$$

Since the highest degree is 3, we are checking to see if this is a cubic spline, i.e. we must check continuity of s, s' and s'' at $x = 1, 2$.

At both points, s is continuous–check that $s(1) = 1$, and $s(2) = 5$ using the piece-wise definitions. However, you can check that $s(x)$ fails to have continuous second derivatives at $x = 1$, so by definition, this cannot be a spline function.

Interpolation using low-degree splines can provide accurate and efficient methods for approximating a function or in some cases, creating a model out of data. Let's investigate linear splines in more depth and then proceed to the special case of cubic spline interpolation. Specifically, a linear spline, or spline of degree 1, is a continuous function whose graph consists of pieces which are all straight lines. Suppose then we are given the values of a function f at the knots

$$a = x_0 < x_1 < \cdots < x_n = b$$

and that we seek the linear spline s which satisfies

$$s(x_k) = f(x_k) \qquad (k = 0, 1, \ldots, n) \tag{6.14}$$

In the knot interval $[x_k, x_{k+1}]$, s must be a polynomial of degree 1 passing through the points $(x_k, f(x_k))$ and $(x_{k+1}, f(x_{k+1}))$. The equation of this line is

$$y = f_k + \frac{f_{k+1} - f_k}{x_{k+1} - x_k}(x - x_k) \tag{6.15}$$

where, as before, we have used f_k to denote $f(x_k)$.

Equation (6.15) for $k = 0, 1, \ldots, n - 1$ can be used to define the required spline function s. In the case of linear spline interpolation, it is straightforward to write down formulas for the coefficients of the various components of the spline function. (unfortunately, this is not so straightforward for all spline functions).

Let s_k be the component of s which applies to the interval $[x_k, x_{k+1}]$. For the linear spline above this gives

$$s_k(x) = a_k + b_k(x - x_k)$$

and from (6.15), we see that

$$a_k = f_k$$

and

$$b_k = \frac{f_{k+1} - f_k}{x_{k+1} - x_k} = f[x_k, x_{k+1}]$$

Example 10 Find the linear spline that interpolates the following data.

x	1	3	4	5.5
$f(x)$	1	27	64	166.375

Here, $x_0 = 1, x_1 = 3, x_2 = 4, x_3 = 5.5$,

$$a_0 = f_0 = 1$$
$$a_1 = f_1 = 27$$
$$a_2 = f_2 = 64$$

and

$$b_0 = f[x_0, x_1] = \frac{27 - 1}{3 - 1} = 13$$

$$b_1 = f[x_1, x_2] = \frac{64 - 27}{4 - 3} = 37$$

$$b_2 = f[x_2, x_3] = \frac{166.375 - 64}{5.5 - 4} = 68.25$$

Therefore

$$s(x) = \begin{cases} 1 + 13(x - 1) = 13x - 12 & 1 \le x \le 3 \\ 27 + 37(x - 3) = 37x - 84 & 3 \le x \le 4 \\ 64 + 68.25(x - 4) = 68.25x - 209 & 4 \le x \le 5.5 \end{cases}$$

The spline is continuous at the interior knots and agrees with the data values at those locations. Note that the linear spline is not defined outside the range of the data. Extrapolation using a linear spline would be especially suspect.

Example 11 Use linear splines to create a line drawing of your hand.

This can be achieved by collecting data (i.e. measurements) that describe the shape of your hand. A rudimentary approach would be simply to place your hand on paper and trace it and then use a ruler to measure points along your hand relative to some location you designate as the origin. Suppose you got n points describing your hand. This gives two data sets: $(i, x_i)_{i=1}^n$ of and $(i, y_i)_{i=1}^n$. Evaluate your linear spline function at a denser set of intermediate points between 1 and n (for example, at increments of 0.05) to obtain the intermediate x and y approximate locations on your hand (let's call those vectors u and v for example). Finally, plot your data along with the intermediate values (Fig. 6.6).

```
>>>  n  =  x.size
>>>  s  =  np.arange(1, n + 1)
>>>  t  =  np.arange(1, n, 0.05)
>>>  u_func  =  interp1d(s, x)
>>>  v_func  =  interp1d(s, y)
>>>  plt.figure(figsize = (3,4))
>>>  plt.plot(x, y, 'k*')
>>>  plt.plot(u_func(t), v_func(t), 'k
    -')
>>>  plt.show()
```

The above example may remind you of connecting the dots when you were a child. When it comes to digital arts applications, a much smoother image would be better. A higher degree polynomial may help. Probably the most commonly used spline functions are *cubic splines* which we will derive in detail below. Following

Fig. 6.6 Linear splines used to recreate a line drawing of a hand using measured locations around the perimeter

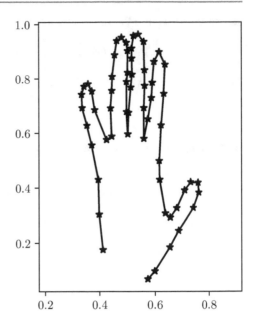

the notation used with linear splines above, the components of a cubic spline can be written as

$$s_k(x) = a_k + b_k(x - x_k) + c_k(x - x_k)^2 + d_k(x - x_k)^3 \qquad (6.16)$$

Now s must satisfy the interpolation conditions

$$s_k(x_k) = f(x_k), \quad s_k(x_{k+1}) = f(x_{k+1}) \qquad (k = 0, 1, \ldots, n-1) \qquad (6.17)$$

which guarantees the continuity of s. The first two derivatives must be continuous giving the following conditions;

$$s_k'(x_{k+1}) = s_{k+1}'(x_{k+1}) \qquad (k = 0, 1, \ldots, n-2) \qquad (6.18)$$
$$s_k''(x_{k+1}) = s_{k+1}''(x_{k+1}) \qquad (k = 0, 1, \ldots, n-2) \qquad (6.19)$$

These requirements give a total of $4n - 2$ equations for the $4n$ coefficients. These equations are linear in the coefficients, so there are two *degrees of freedom*. There are several ways of using these, but we consider just the *natural cubic spline*, described more as the derivations unfold here.

First, the interpolation conditions $s_k(x_k) = f(x_k)$ yield

$$a_k = f_k \qquad (k = 0, 1, \ldots, n-1) \qquad (6.20)$$

Substituting this into the remaining interpolation conditions we get

$$b_k (x_{k+1} - x_k) + c_k (x_{k+1} - x_k)^2 + d_k (x_{k+1} - x_k)^3 = f_{k+1} - f_k \qquad (6.21)$$

For the sake of notation, denote the steplengths by h_k, so that

$$h_k = x_{k+1} - x_k.$$

Substituting this into (6.21) and dividing by h_k, we now have, for $k = 0, 1, \ldots,$
$n - 1$,

$$b_k + c_k h_k + d_k h_k^2 = \frac{f_{k+1} - f_k}{h_k} = f\left[x_k, x_{k+1}\right] = \delta_k. \qquad (6.22)$$

Next, substituting into (6.18) and (6.19) gives

$$b_k + 2c_k h_k + 3d_k h_k^2 = b_{k+1} \qquad (k = 0, 1, \ldots, n - 2) \qquad (6.23)$$

and

$$2c_k + 6d_k h_k = 2c_{k+1} \qquad (k = 0, 1, \ldots, n - 2)$$

This last equation gives

$$d_k = \frac{c_{k+1} - c_k}{3h_k} \qquad (6.24)$$

and substituting (6.24) into (6.22) yields

$$b_k = \delta_k - c_k h_k - \frac{(c_{k+1} - c_k) h_k}{3}$$

$$= \delta_k - \frac{h_k}{3} (c_{k+1} + 2c_k) \qquad (6.25)$$

To this end, if the coefficients c_k can be determined, then Eqs. (6.24) and (6.25) can be used to complete the definition of the spline components. Substituting for b_k, b_{k+1}, and d_k in (6.23), we arrive at a linear system of equations for the coefficients c_k, as follows:

$$\delta_k - \frac{h_k}{3} (c_{k+1} + 2c_k) + 2c_k h_k + (c_{k+1} - c_k) h_k = \delta_{k+1} - \frac{h_{k+1}}{3} (c_{k+2} + 2c_{k+1})$$

Collecting terms, we obtain

$$h_k c_k + 2 (h_k + h_{k+1}) c_{k+1} + h_{k+1} c_{k+2} = 3 (\delta_{k+1} - \delta_k) \qquad (6.26)$$

for $k = 0, 1, \ldots, n - 3$. The result is a *tridiagonal* system of linear equations for the unknown coefficients c_k (which we know how to solve efficiently from our previous chapter). The system (6.26) however has only $n - 2$ equations for the n unknowns. This is where the two degrees of freedom discussed earlier will be used.

Definition 20 *The* natural cubic spline *is defined by imposing the additional conditions*

$$s''(a) = s''(b) = 0$$

This requirement implies that the spline function continues with straight lines outside the interval $[a, b]$ while maintaining its smoothness. This mimics the behavior of the physical spline beyond the extreme knots.

For the natural cubic spline, this implies that

$$s''(a) = s_0''(x_0) = 2c_0 = 0$$

while

$$s''(b) = s_{n-1}''(x_n) = 2c_{n-1} + 6d_{n-1}h_{n-1} = 0$$

Introducing the spurious coefficient $c_n = 0$, this last equation becomes

$$d_{n-1} = \frac{c_n - c_{n-1}}{3h_{n-1}}$$

which is to say that (6.24) remains valid for $k = n - 1$. It also extends the validity of (6.26) to $k = n - 2$.

The full tridiagonal system is

$$H \begin{bmatrix} c_1 \\ c_2 \\ \vdots \\ c_{n-1} \end{bmatrix} = 3 \begin{bmatrix} \delta_1 - \delta_0 \\ \delta_2 - \delta_1 \\ \vdots \\ \delta_{n-1} - \delta_{n-2} \end{bmatrix}$$

where H is the tridiagonal matrix

$$\begin{bmatrix} 2(h_0 + h_1) & h_1 & & & & \\ h_1 & 2(h_1 + h_2) & h_2 & & & \\ & h_2 & 2(h_2 + h_3) & h_3 & & \\ & & \ddots & \ddots & \ddots & \\ & & & h_{n-3} & 2(h_{n-3} + h_{n-2}) & h_{n-2} \\ & & & & h_{n-2} & 2(h_{n-2} + h_{n-1}) \end{bmatrix}$$

Now there are $n - 2$ equations in the $n - 2$ remaining unknowns.

Before looking at the implementation of cubic spline interpolation within an algorithmic setting, an example can demonstrate how all the mathematics interplays to build the spline.

Example 12 Find the natural cubic spline which fits the data

$$\begin{array}{c|ccccc} x & 1 & 4 & 9 & 16 & 25 \\ \hline f(x) & 1 & 2 & 3 & 4 & 5 \end{array}$$

Since there are five knots, $n = 4$ and we should obtain a 3×3 tridiagonal system of linear equations for the unknown coefficients c_1, c_2, c_3.

Recall that $a_k = f_k$ for $k = 0, 1, 2, 3$ so

$$a_0 = 1, a_1 = 2, a_2 = 3, a_3 = 4$$

and the steplengths are $h_0 = 3, h_1 = 5, h_2 = 7, h_3 = 9$. The divided differences computations give:

$$\delta_0 = 1/3, \delta_1 = 1/5, \delta_2 = 1/7, \delta_3 = 1/9$$

The tridiagonal system is then

$$2(3+5)c_1 + 5c_2 = 3\left(\frac{1}{5} - \frac{1}{3}\right) = -0.4$$

$$5c_1 + 2(5+7)c_2 + 7c_3 = 3\left(\frac{1}{7} - \frac{1}{5}\right) = -0.1714$$

$$7c_2 + 2(7+9)c_3 = 3\left(\frac{1}{9} - \frac{1}{7}\right) = -0.0952$$

In matrix terms:

$$\begin{bmatrix} 16 & 5 & 0 \\ 5 & 24 & 7 \\ 0 & 7 & 32 \end{bmatrix} \begin{bmatrix} c_1 \\ c_2 \\ c_3 \end{bmatrix} = \begin{bmatrix} -0.4000 \\ -0.1714 \\ -0.0952 \end{bmatrix}$$

The solution to this system is

$$c_1 = -2.4618 \times 10^{-2}, \quad c_2 = -1.2233 \times 10^{-3}, \quad c_3 = -2.7074 \times 10^{-3}$$

Since $c_0 = c_4 = 0$ (remember (6.24)), we get

$$d_0 = \frac{c_1 - c_0}{3h_0} = -2.7353 \times 10^{-3}$$

$$d_1 = 1.5595 \times 10^{-3}$$

$$d_2 = -7.0671 \times 10^{-7}$$

$$d_3 = 1.0031 \times 10^{-4}$$

Similarly, substituting into (6.25), we get

$$b_0 = \delta_0 - \frac{h_0}{3}(c_1 + 2c_0) = 3.5795 \times 10^{-1}$$
$$b_1 = 2.8410 \times 10^{-1}$$
$$b_2 = 1.5489 \times 10^{-1}$$
$$b_3 = 1.2736 \times 10^{-1}$$

Putting it all together, components of the natural cubic interpolation spline are then defined by using the appropriate values from above in

$$s_k(x) = a_k + b_k(x - x_k) + c_k(x - x_k)^2 + d_k(x - x_k)^3$$

So for $x = 3 \in [1, 4]$, we use $k = 0$ and $x - x_0 = 2$ so that

$$s(3) = 1 + \left(3.5795 \times 10^{-1}\right)(2) + 0(2)^2 + \left(-2.7353 \times 10^{-3}\right)(2)^3 = 1.6940$$

Likewise, for $x = 7 \in [4, 9]$, we use $k = 1$, $x - x_1 = 3$ and obtain

$$s(7) = 2 + \left(2.8410 \times 10^{-1}\right)(3) + \left(-2.4617 \times 10^{-2}\right)(3)^2 + \left(1.5595 \times 10^{-3}\right)(3)^3 = 2.6728$$

The absolute errors in the example values above are approximately 3.8051×10^{-2} and 2.7049×10^{-2} which are of a similar order of magnitude to those that would be expected from using local cubic divided difference interpolation. (The data in Example 12 is from the square-root function.) This is to be expected since the error in natural cubic spline interpolation is again of order h^4 as it is for the cubic polynomial interpolation. The proof of this result for splines is much more difficult however. It is left to subsequent courses and more advanced texts.

Typically, the natural cubic spline will have better smoothness properties because, although the approximating polynomial pieces are local, they are affected by the distant data. In order to get a good match to a particular curve, we must be careful about the placement of the knots. We shall consider this further, though briefly, after discussing the implementation of natural cubic spline interpolation.

The function cspline first computes the coefficients using the equations derived earlier. The tridiagonal system is solved here using NumPy's linear equation solver numpy.solve. We shall discuss techniques for solving such systems in detail in Chap. 7. For now we shall content ourselves with the "black box" provided with Python/NumPy.

Program Python function for cubic spline interpolation

```python
import numpy as np

def cspline(knots, data, x):
    """
    Performs natural cubic spline interpolation at
    points 'x' based on 'knots' with values given
    as 'data'.
    """

    N = np.size(knots) - 1
    P = np.size(x)
    h = np.diff(knots)
    D = np.diff(data) / h
    dD3 = 3 * np.diff(D)
    a = data[:N]

    # Genereate tri-diagonal system
    H = (np.diag(2 * (h[:-1] + h[1:]))
        + np.diag(h[1:-1], 1)
        + np.diag(h[1:-1], -1))
    c = np.zeros(N+1)
    c[1:N] = np.linalg.solve(H, dD3)
    b = D - h * (c[1:] + 2 * c[:-1]) / 3
    d = (c[1:] - c[:-1]) / (3 * h)

    # Evaluate spline at x
    s = np.empty(P)
    for i in range(P):
        indices = np.argwhere(x[i] > knots)
        if indices.size > 0:
            k = indices.flat[-1]
            # x[i] lies in the knot interval
            # (knots[k], knots[k+1]]
        elif x[i] == knots[0]:
            k = 0  # Special case: include x
                   # in first interval, i.e.
                   # in [knots[0], knots[1]]
        else:
            raise ValueError('cspline does not\
                support extrapolation - values\
                outside the knot intervals.')
        z = x[i] - knots[k]
        s[i] = a[k] + z * (b[k] + z * (c[k]
                                  + z * d[k]))

    return s
```

Note the use of array arithmetic operations in generating the remaining coefficients once the c's have been computed. Note, too, the use of np.diag to initialize the matrix elements.

Fig. 6.7 Error in cubic spline interpolation

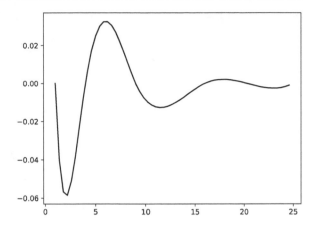

This function assumes that the arguments x are all in the interval spanned by the knots. It would evaluate the spline outside this interval by using the first component for any x-value to the left, and the last component for values to the right of the span of the knots. It is not difficult to modify the code to extend the spline with straight line components beyond this interval but we do not trouble with this here.

Example 13 Plot the error function $s(x) - \sqrt{x}$ for the data of Example 12.

We can use the previously created Python function cspline with the following commands to generate the graph in Fig. 6.7.

```
>>> import numpy as np
>>> from matplotlib import pyplot as plt
>>> from prg_cubicspline import cspline

>>> X = np.array([1, 4, 9, 16, 25])
>>> Y = np.array([1, 2, 3, 4, 5])
>>> x = np.arange(1, 25, .4)
>>> y = cspline(X, Y, x)
>>> plt.plot(x, y - np.sqrt(x))
>>> plt.show()
```

Here the vectors X, Y are the knots and the corresponding data values, x is a vector of points at which to evaluate the natural cubic spline interpolating this data and y is the corresponding vector of spline values.

We see that this error function varies quite smoothly, as we hoped. Also we note that the most severe error occurs very near $x = 3$ which was the first point we used in the earlier example. The evidence presented there was representative of the overall performance (Fig. 6.7).

In motivating the study of spline interpolation, we used an example of local cubic divided difference interpolation where the resulting piecewise polynomial had a

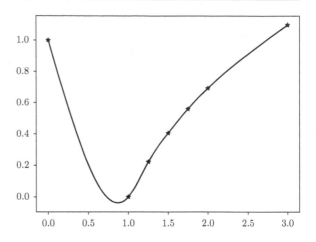

Fig. 6.8 Interpolation with a cubic spline

discontinuity. See Fig. 6.5. The corresponding cubic spline interpolant is plotted in Fig. 6.8. This was generated using the Python commands

```
>>> X = np.concatenate(((0,),
        np.arange(1, 2.25, .25), (3,)))
>>> Y = np.concatenate(((1,), np.log(X[1:])))
>>> x = np.linspace(X[0], X[-1], 200)
>>> y = cspline(X, Y, x)
>>> plt.plot(x, y, 'k-')
>>> plt.plot(X, Y, 'k*')
>>> plt.show()
```

It is immediately apparent that the cubic spline has handled the changes in this curve much more readily. The main cause of difficulty for the local polynomial interpolation used earlier is the fact that the function has a minimum and an inflection point very close together – and the inflection point appears to be very close to the discontinuity of the local divided difference interpolation function.

In general, we want to place more nodes in regions where the function is changing rapidly (especially in its first derivative). We shall consider briefly the question of knot placement for cubic spline interpolation in the next example.

Example 14 Use cubic spline interpolation for data from the function $1/(1 + x^2)$ with eleven knots in $[-5, 5]$.

First, for equally spaced knots, we can use the Python commands

```
>>> X = np.arange(-5, 6)
>>> Y = runge(X)
>>> x = np.arange(-5, 5.05, .05)
>>> y = cspline(X, Y, x)
>>> plt.plot(X, Y, 'k*', x,
```

```
            runge(x), 'k-', x, y, 'k-')
>>> plt.show()
>>> plt.figure()
>>> plt.plot(x, runge(x) - y, 'k-')
>>> plt.show()
```

to obtain the following two plots in Figs. 6.9 and 6.10. In Fig. 6.9, we see that the graphs of Runge and its cubic spline interpolant using integer knots in $[-5, 5]$ are essentially indistinguishable. The error function in Fig. 6.10 shows that the fit is especially good near the ends of the range (where polynomial interpolation failed). The errors are largest either side of ± 1 which is where the inflection points are, and where the gradient is changing rapidly. To get a more uniform fit we would probably want to place more of the knots in this region.

In Fig. 6.11 we see the effect on the error function of using knots at

$$-5, -3.5, -2.5, -1.5, -0.75, 0, 0.75, 1.5, 2.5, 3.5, 5$$

Fig. 6.9 Interpolation of the Runge function with a cubic spline

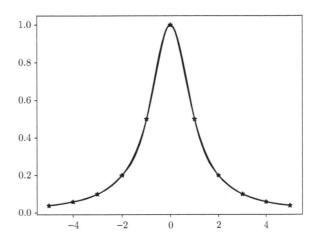

Fig. 6.10 Error of the interpolation from Fig. 6.9

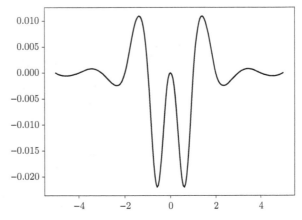

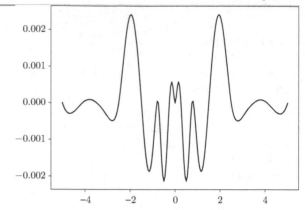

Fig. 6.11 Interpolation error with new knots

There are still just eleven knots but the magnitude of the maximum error has been reduced by an approximate factor of 10.

We have seen that natural cubic splines can be used to obtain smooth interpolating functions even for data where polynomial interpolation cannot reproduce the correct behavior.

Exercises

1. Which of the following functions are splines of degree 1, 2, or 3?

$$
\text{(a) } s(x) = \begin{cases} x & 0 \le x \le 1 \\ 2x - 1 & 1 \le x \le 2 \\ x + 2 & 2 \le x \le 4 \end{cases}
$$

$$
\text{(b) } s(x) = \begin{cases} 2 - x & 0 \le x \le 1 \\ 2x - 1 & 1 \le x \le 2 \\ x + 1 & 2 \le x \le 4 \end{cases}
$$

$$
\text{(c) } s(x) = \begin{cases} x^2 & -1 \le x \le 1 \\ 2x^2 - 2x + 1 & 1 \le x \le 2 \\ 3x^2 - 6x + 5 & 2 \le x \le 3 \end{cases}
$$

$$
\text{(d) } s(x) = \begin{cases} x^3 + 1 & 0 \le x \le 1 \\ 2x^2 - 2x + 2 & 1 \le x \le 2 \\ 3x^2 - 6x + 3 & 2 \le x \le 4 \end{cases}
$$

$$
\text{(e) } s(x) = \begin{cases} x & -1 \le x \le 0 \\ x + x^3 & 0 \le x \le 1 \\ 1 - 2x + 3x^2 & 1 \le x \le 4 \end{cases}
$$

2. Find the linear spline which interpolates the data
   What are its values at 2, 3.5 and 4.5?

x	0	1	3	4	6
$f(x)$	5	4	3	2	1

3. Write a program to compute the linear spline interpolating a given set of data. Use it to plot the linear spline interpolant for the data in Exercise 2. Also plot the linear interpolating spline for the function $1/\left(1+x^2\right)$ using knots $-5, -4, \ldots, 5$. (If *all* we want is the plot, there is a simpler way, of course. Just plot(X,Y) where X and Y are the vectors of knots and their corresponding data values.)

4. Find the natural cubic interpolation spline for the data

x	1	2	3	4	5
$\ln x$	0.0000	0.6931	1.0986	1.3863	1.6094

- **Hints**: As usual, the a-coefficients are just the data values:

$$a_0 = 0.0000 \quad a_1 = 0.6931 \quad a_2 = 1.0986 \quad a_3 = 1.3863.$$

You should obtain the tridiagonal system for c_1, c_2, c_3:

$$\begin{bmatrix} 4 & 1 & 0 \\ 1 & 4 & 1 \\ 0 & 1 & 4 \end{bmatrix} \begin{bmatrix} c_1 \\ c_2 \\ c_3 \end{bmatrix} = \begin{bmatrix} -0.8628 \\ -0.3534 \\ -0.1938 \end{bmatrix}$$

5. Compute the natural cubic spline interpolant for data from the function $\sqrt{x}\exp\left(-x\right)$ using knots at the integers in $[0, 8]$. Compare the graph of this spline with that of $\sqrt{x}\exp\left(-x\right)$.

6. We can continue a natural cubic spline outside the span of the knots with straight lines while preserving the continuity of the first two derivatives. Show that to do this we can use $s_{-1}(x) = a_0 + b_0(x - x_0)$ and $s_n(x) = a_n + b_n(x - x_n)$ where

$$a_n = f(x_n), \text{ and}$$
$$b_n = \frac{c_{n-1}h_{n-1}}{3} + \delta_{n-1}$$

7. Modify the program for natural cubic spline interpolation to include straight line extensions outside the span of the knots. Use this modified code to plot the natural cubic spline interpolant to $1/\left(1+x^2\right)$ over the interval $[-6, 6]$ using knots $-5, -4, \ldots, 5$.

8. Repeat Exercise 5, using 17 equally spaced knots, 0:0.5:8.
9. Try to find the best set of 11 knots for natural cubic spline interpolation to $1/(1+x^2)$ over $[-5,5]$.
10. Create a drawing of your hand similar to the approach in Example 11, but using cubic splines. Take it further–find the area of your hand by integrating the polynomial exactly and then comparing your result to what you would get with composite Simpson's rule. Explain what you see.
11. Choose a picture to represent with splines. Use spline interpolation to reproduce your picture as a drawing. You will need to choose and measure data points relative to a convenient set of axes. Use different splines to approximate different curves in the picture and plot them all on the same set axes. For more vertically-oriented curves, you can simply interchange the roles of the variables in your call to the spline function and then reverse the order in your plot command. Include these aspects in your report:

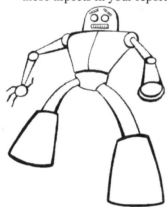

(a) Original picture

(b) Reproduction using splines (made by student T. Penderghest)

- Discuss your choice of original picture (line, photograph, cartoon, painting) and the implications of your choice for this project.
- Discuss your rationale for how many points to use and where to locate the points.
- Describe which splines you chose to use for various features in the image. If you tried something that did not work as well as your final choice, include that information.
- Discuss how the reproduction qualitatively compares (visually) to the original. What aspects of the original have you captured? Which have you missed or omitted?

12. Repeat the above exercise but this time use a picture of yourself to create a line-drawing self portrait.

6.5   Conclusions and Connections: Interpolation

What have you learned in this chapter? And where does it lead?

The chapter began with polynomial interpolation in which we seek a polynomial of specified degree that agrees with given data. The first approach was the "obvious" one of using our knowledge of solving linear systems of equations to find the Lagrange interpolation polynomial by solving the Vandermonde system for the coefficients. However that is both inefficient and because of ill-conditioning subject to computational error.

The use of the Lagrange basis polynomials is equivalent to transforming that system to a diagonal form, and is a more practical approach. Even so the addition of new data points is cumbersome–and this is a very real issue when for example the data comes from a time series and new observations become available–such as adding the temperature at 11:00 to a sequence of temperature readings at 10:00, 10:15, 10:30 and 10:45 in forecasting the temperature at 11:15. (Think of the Weather Channel app preparing its future radar to forecast the progress on a winter storm. They would certainly want to use the most recent true data to update those estimates.)

Difference schemes allow a more efficient use of the data, including adding new data points. These divided difference approaches have a very long history–which is why they bear Newton's name, of course–but they retain their importance today. Part of the significance of difference representations lies in the ability to recenter the data so that local data assumes greater importance relative to more distant data points. It also allows us to halt computation (effectively reducing the degree of the polynomial) either when no further improvement results from using more data, or when all data within some reasonable radius of the point of interest has been exhausted. That is the interpolation can be local.

Traditional polynomial, or even spline, interpolation is not always appropriate although it provides initial insight into a deep field of applied mathematics. In some scenarios, other methods will give better results, and it is helpful to draw a distinction between the approximation of a smooth function whose tabulated values are known very accurately, and the situation where the data itself is subject to error. (Most practical problems have some aspects of both of these.) In the former situation, polynomial approximation may be applied with advantage over a wide interval, sometimes using polynomials of moderate or high degree. In the latter case, it is invariably desirable to use lower degree polynomials over restricted ranges. One important technique is approximation by cubic splines in which the function is represented by different cubic polynomials in different intervals.

Spline interpolation, can be viewed as a very special form of local polynomial interpolation. Here the basic idea is to use low degree polynomials which connect as smoothly as possible as we move through the data. The example we focused on was cubic spline interpolation where the resulting function retains continuity in its slope and curvature at each data point. (We enforce those through continuity of the first and second derivatives.) Spline interpolation was an important tool in creating the early digitally animated movies such as Toy Story. The ability to have visually smooth curves that do not deteriorate under scaling for example is critical to creating good images. In the context of animation, splines have been largely superseded in recent years by subdivision surfaces. These are similarly smooth curves (or surfaces in higher dimension) where certain points act as control points which you can envision as elastic connections to the curve rather than fixed interpolation points. This is one of many potential research topics that has its origins in the methods of this chapter.

6.6 Python Interpolation Functions

Python has several functions for interpolation available in SciPy. These include polynomial as well as spline interpolation.

scipy.interpolate.interp1d is a joint interface to several different types of interpolation in SciPy. It takes as a minimum two arguments; an array of nodes and an array of corresponding data values. Additionally, we can set the optional argument kind to specify the kind of interpolation to use:

'linear' default value – performs linear interpolation between the data values.
'nearest' interpolates intermediate values to the nearest data value, i.e. piece-wise constant interpolation.
'zero' zero-degree spline interpolation.
'slinear' first-degree (linear) spline interpolation.
'quadratic' second-degree spline interpolation.
'cubic' third-degree spline interpolation.

The interp1d function returns a function itself. The returned function can be used to evaluate the interpolation at an array of points supplied to it.

Example 15 We can use the following data to interpolate a set of points using SciPy's interp1d function

```
X = np.concatenate (((0 ,) , np.arange (1 , 2.25 , .25) ,
                     (3 ,) ))
Y = np.concatenate (((1 ,) , np.log (X [1:]) ))
```

The interpolating functions are obtained using

```
y1 = interp1d (X, Y)
y2 = interp1d (X, Y, 'nearest')
y3 = interp1d (X, Y, 'zero')
y4 = interp1d (X, Y, 'slinear')
y5 = interp1d (X, Y, 'quadratic')
y6 = interp1d (X, Y, 'cubic')
```

The interpolation y1 uses the default argument 'linear'. So far we have the functions y1 to y6 available for evaluating the interpolation at arbitrary points. We can do this directly in plotting the interpolations

```
x = np.arange (0, 3, 0.02)

plt.plot (x, y1 (x), 'k')
plt.plot (x, y2 (x), 'k')
plt.plot (x, y3 (x), 'k')
plt.plot (x, y4 (x), 'k')
plt.plot (x, y5 (x), 'k')
plt.plot (x, y6 (x), 'k')
plt.show ()
```

which are shown in Fig. 6.12.

We see that the interpolations obtained using 'linear' (default) and 'slinear' coincide.

SciPy similarly has the functions interp2d and interpn for interpolation in 2 dimensions and in n dimensions, respectively. In addition, the scipy.interpolate module contains several additional functions and classes for performing various types of interpolation. We advise you to explore the SciPy documentation for further details.

numpy.polyfit was introduced in Sect. 4.8.2. It is intended for *least squares fitting* of polynomials to data. As such, it can also be used to compute an interpolating

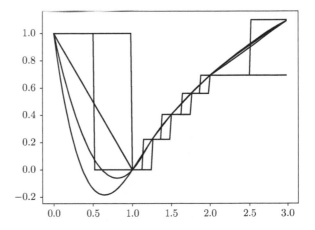

Fig. 6.12 Six different interpolation types using SciPy's interp1d

polynomial for a set of nodes and corresponding function values. See Sect. 4.8.2 for an example.

We can use numpy.polyfit configured for degree one less than the number of nodes to obtain full polynomial interpolation in Python. The resulting polynomial can be evaluated using the function numpy.polyval.

Differential Equations

7

7.1 Introduction and Euler's Method

Finally we come to the topic of differential equations, a widely used mathematical approach to understanding the world around us. Observing that the world around us is changing and we'd like to make predictions about how it is changing leads us to the notion of derivatives, which represent rates of change. Differential equations involve an unknown function that must satisfy a certain relationship between its derivatives.

The Newton's law of cooling example used earlier was actually based on the following assumption, that the "rate of change of the temperature of an object is proportional to the difference between the ambient environment and the temperature of the object". Mathematically (and using our previous notation with the coffee model), this implies

$$T'(t) = k(T_E - T(t)).$$

The coffee model used earlier was found by solving this differential equation analytically.

The application of differential equations in modeling covers a broad range of possibilities from population dynamics, to the spread of diseases, to the growth of a tumor. Note, these are different than partial differential equations, which describe phenomena changing in space and time (you saw a flavor for this with the heat equation on a wire and image blurring examples).

To this end, we focus here on the basic ideas behind the numerical solution of differential equations, concentrating primarily on the solution of first-order *initial-value problems*. These have the basic form

$$y' = f(x, y); \qquad y(x_0) = y_0 \tag{7.1}$$

Later in the chapter both higher-order initial value problems (or systems of them) and the solution of *boundary value problems* (where the conditions are specified

© Springer International Publishing AG, part of Springer Nature 2018   229
P. R. Turner et al., *Applied Scientific Computing*, Texts in Computer Science,
https://doi.org/10.1007/978-3-319-89575-8_7

at two distinct points) are also considered briefly. A typical second-order two-point boundary value problem has the form

$$y'' = f\left(x, y, y'\right); \qquad y\left(a\right) = y_a, y\left(b\right) = y_b \qquad (7.2)$$

Taylor series are the basis for many of the methods to follow. That is, some methods can be derived by considering

$$y\left(x_1\right) = y\left(x_0 + h_0\right) = y_0 + h_0 y'\left(x_0\right) + \frac{h_0^2}{2} y''(x_0) + \cdots \qquad (7.3)$$

for some steplength h_0. Truncating the series after a certain number of terms or approximating terms using other function values are some ways to arrive at numerical schemes to approximate the solution at x_1. The process can then be repeated for subsequent steps. An approximate value $y_1 \approx y\left(x_1\right)$ is used with a steplength h_1 to obtain an approximation $y_2 \approx y\left(x_2\right)$, and so on throughout the domain of interest.

Although Taylor series provides one way to derive many of the methods discussed here, there are other approaches. Many of the methods can also be viewed as applications of simple numerical integration rules to the integration of the function $f\left(x, y\left(x\right)\right)$. The simplest method – from either of these viewpoints – is *Euler's method* for which there is also an insightful graphical explanation. We should note, you already got a flavor for this in the section on numerical calculus, but we present here from a beginning viewpoint. To get an initial understanding, we'll consider Euler's method as a graphical technique using an example. Figure 6.1 shows a slope field for the differential equation

$$y' = (6x^2 - 1)y \qquad (7.4)$$

which, with the initial condition $y\left(0\right) = 1$, we shall use as a basic example for much of this chapter. This particular initial value problem

$$y' = (6x^2 - 1)y; \qquad y\left(0\right) = 1 \qquad (7.5)$$

is easily solved by standard techniques. However that allows us to compare our computed solution with the true one to analyze the errors.

The *slope field* (or direction field) consists of short line segments whose slopes are the slopes of solutions to (7.4) passing through those points. These slopes are computed by evaluating the right-hand side of the differential equation at a grid of points (x, y). The line segments are therefore tangent lines to solutions of the differential equation at the various points.

Using the slope field it is also easy to see the basic shape of the solution through any particular point. Try tracing a curve starting at $(0, 1)$ in Fig. 7.1 following the slopes indicated and you will quickly obtain a good sketch of the solution to the initial value problem (7.5).

Fig. 7.1 Slope field for
Eq. (7.4)

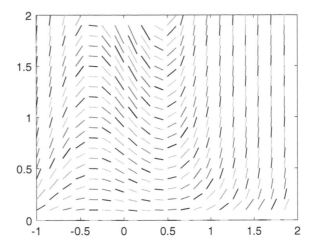

Following a tangent line for a short distance gives an approximation to the solution at a nearby point. The tangent line at that point can then be used for the next step in the solution process. That is the conceptual idea behind Euler's method.

As mentioned above, Euler's method can be derived algebraically from a Taylor approximation. Consider for these derivations, a fixed steplength h and let $x_0 = 0$ and $y_0 = 1$ as in (7.5). We denote $x_0 + kh$ by x_k for $k = 0, 1, \ldots$ A first-order Taylor approximation to $y(x_1)$ is

$$y(x_1) = y(x_0 + h) \approx y_0 + hy'(x_0) = y_0 + hf(x_0, y_0)$$

Let the approximation to $y(x_1)$ be denoted by y_1 so that

$$y_1 = y_0 + hf(x_0, y_0) \tag{7.6}$$

is the first step of Euler's method.

Continuing in this manner, gives the general Euler step:

$$y_{k+1} = y_k + hf(x_k, y_k) \tag{7.7}$$

In terms of notation, keep in mind that apart from y_0, we use y_k to represent our approximation to the solution value $y(x_k)$. This is in contrast to our notation for interpolation and integration where we often used $f_k = f(x_k)$.

Also note that (7.7) is not quite the desired generalization of (7.6). In the first step (7.6), the exact values are used in the right-hand side. In subsequent steps, previously computed approximate values are used.

The notion that many methods for differential equations can be derived in terms of the approximate integration of the right-hand side of the differential equation

(7.1) is worth revisiting. To see this, consider the first step of the solution. By the fundamental theorem of calculus, we have

$$y(x_1) - y(x_0) = \int_{x_0}^{x_1} f(x, y(x)) \, dx. \tag{7.8}$$

Approximating this integral with the left-hand end-point rule, we obtain

$$y(x_1) - y(x_0) \approx (x_1 - x_0) f(x_0, y(x_0)) = hf(x_0, y(x_0))$$

which is equivalent to the approximation (7.6) derived above for Euler's method.

This suggests that alternative techniques could be based on better numerical integration rules such as the mid-point rule or Simpson's rule. This is not completely straightforward since such formulas would require knowledge of the solution at points where it is not yet known. This is discussed more later.

The Python implementation of Euler's method is below. The inputs required are the function f, the interval $[a, b]$ over which the solution is sought, the initial value $y_0 = y(a)$, and the number of steps to be used. The following code can be used.

Program   Python implementation of Euler's method.

```
import numpy as np

def euler(fcn, a, b, y0, N):
    """
    Function for generating an Euler solution to
    y' = f(x,y) in 'N' steps with initial
    condition y[a] = y0.

    """

    h = (b - a) / N
    x = a + np.arange(N+1) * h
    y = np.zeros(x.size)
    y[0] = y0
    for k in range(N):
        y[k+1] = y[k] + h * fcn(x[k], y[k])

    return (x, y)
```

The output here consists of a table of values x_k, y_k.

Example 1 Apply Euler's method to the solution of the initial value problem (7.5) $y' = (6x^2 - 1)y$; $y(0) = 1$. Begin with $N = 4$ steps and repeatedly double this number of steps up to 256.

Applying the program above with $N = 4$ we get

x	y
0	1.0000
0.25	0.7500
0.5	0.6328
0.75	0.7119
1.0	1.1346

It is often useful to check computer code with some hand calculations. We step through Euler's method here by hand to get a better idea of how it works. With $x_0 = 0$, $y_0 = 1$, $N = 4$, and $h = 1/4$, we obtain

$$y_1 = y_0 + \frac{1}{4}(6x_0^2 - 1)y_0 = \frac{3}{4}$$

Then with $x_1 = 1/4$, $y_1 = \frac{3}{4}$:

$$y_2 = y_1 + \frac{1}{4}(6x_1^2 - 1)y_1 = \frac{3}{4} + \frac{1}{4}\left(6\left(\frac{1}{4}\right)^2 - 1\right)\frac{3}{4} = 0.6328$$

and continuing in this way we get

$$y_3 = y_2 + \frac{1}{4}(6x_2^2 - 1)y_2 = 0.6328 + \frac{1}{4}\left(6\left(\frac{1}{2}\right)^2 - 1\right)0.6328 = 0.7119$$

$$y_4 = y_3 + \frac{1}{4}3x_3^2 y_3 = 0.7119 + \frac{1}{4}\left(6\left(\frac{3}{4}\right)^2 - 1\right)0.7119 = 1.1346$$

Similar tables, with more entries are generated for the other values of N. The values for $y(1)$ obtained, and their errors are tabulated below. (Note that the true solution of (7.5) is $y = \exp\left(2x^3 - x\right)$ so that $y(1) = e$.)

N	4	8	16	32	64	128	256
y_N	1.1346	1.6357	2.0527	2.3420	2.5169	2.6139	2.6651
$|Error|$	1.5837	1.0826	0.6656	0.3763	0.2014	0.1044	0.0532

Although it is slow, the errors are decreasing. The solutions are plotted, along with the true solution in Fig. 7.2.

As seen in the numerical calculus section, the ratio of successive errors can be used to verify that code is working properly by comparing values to what the truncation error theory predicts. Here the ratios of the successive final errors in Example 1 are approximately 0.6836, 0.6148, 0.5352, 0.5183, and 0.5095 which seem to be settling

Fig. 7.2 Euler's method as N increases from 4 to 256. Note how the approximations approach the true solution as N increases

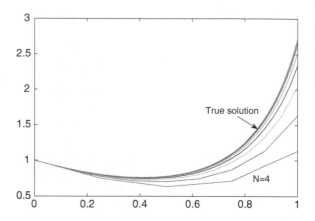

down close to 1/2 which is the same factor by which the steplength h is reduced. This suggests that the overall error in Euler's is first order, i.e. the error is $\mathcal{O}(h)$.

Let's take a closer look at the truncation error analysis. The successive approximate solution points are generated by following the tangents to the different solution curves through these points. At each step the approximate solution follows the tangent line. At the next step, it follows the tangent to the solution to the original differential equation that passes through this *erroneous* point. Although all the solution curves have similar behavior, those below the desired solution are all less steep at corresponding x-values and the convex nature of the curves implies that the tangent will lie below the curve resulting in further growth in the error.

So, there are two components to this error; one is a contribution due to the straight line approximation used at each step. Second, there is a significant contribution resulting from the accumulation of errors from previous steps. Their effect is that the linear approximation is not tangent to the required solution but to another solution of the differential equation passing through the current point. These two types of error contributions are called the *local* and *global* truncation errors.

The situation is illustrated in Fig. 7.3 for the second step of Euler's method. In this case, the local contribution to the error at step 2 is relatively small, with a much larger contribution coming from the effect of the (local and global) error in the first step.

For the first step, we have

$$y_1 = y_0 + hf(x_0, y_0)$$

while Taylor's theorem gives us

$$y(x_1) = y_0 + hy_0' + \frac{h^2}{2}y''(\xi) = y_0 + hf(x_0, y_0) + \frac{h^2}{2}y''(\xi)$$

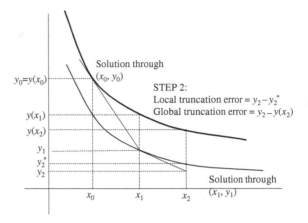

Fig. 7.3 Illustration of local and global truncation error after two steps

It follows that the local truncation error in the approximation is

$$|y_1 - y(x_1)| = \frac{h^2}{2}|y''(\xi)| \leq Kh^2$$

where K is a bound on the second derivative of the solution. A similar local truncation error will occur at each step of the solution process.

We see that the local truncation error for Euler's method is $O(h^2)$ yet we observed that the global truncation error appears to be just $O(h)$. By the time we have computed our estimate y_N of $y(b)$, we have committed N of these truncation errors from which we may conclude that the global truncation error has the form

$$E = NKh^2 = K(b-a)h$$

since $Nh = b - a$. Thus the global truncation error is indeed of order h. The stated error is the leading contribution to the rigorously defined global truncation error which is of order h. (The rigorous derivation is more complicated than is appropriate here.)

In addition to the truncation error theory presented here, in practice there will also be round-off error, which will tend to accumulate. Those effects can be analyzed in a manner similar to that for the global truncation error. The result is that the global roundoff error *bound* is *inversely* proportional to the steplength h. As with numerical differentiation earlier in this text, this places a restriction on the accuracy that can be achieved. However, for systems working with IEEE double precision arithmetic, the effect of roundoff error is typically *much* smaller than the global truncation error unless we are seeking very high accuracy and therefore using very small steplengths.

Euler's method is straightforward to implement. Unfortunately, it is not always going to be sufficient to generate accurate solutions to differential equations. More accurate methods are developed in the next sections that build on similar ideas presented above.

Exercises

Exercises 1–6, and 8 concern the differential equation

$$y' = 2yx - y$$

1. Find the general solution of the differential equation.
2. For the initial condition $y(0) = 3$, use Euler's method with steps $h = 1, 1/2$ and $1/4$ to approximate $y(1)$.
3. Use Euler's method to solve the initial value problem of Exercise 2 over $[0, 2]$ with $N = 10, 20, 50, 100, 200$ steps.
4. Tabulate the errors in the approximate values of $y(2)$ in Exercise 3. Verify that their errors appear to be $O(h)$.
5. Using negative steplengths solve the initial value problem in Exercise 2 over $[-2, 0]$ to obtain a graph of its solution over $[-2, 2]$.
6. Repeat Exercise 5 for different initial conditions: $y(0) = 1, 2, 3, 4, 5$ and plot the solutions on the same axes.
7. Use Euler's method with steplengths $h = 10^{-k}$ for $k = 1, 2, 3$ to solve the initial value problem $y' = x + y^2$ with $y(0) = 0$ on $[0, 1]$. Tabulate the results for $x = 0, 0.1, 0.2, \ldots, 1$ and graph the solutions.
8. (More challenging) Use Euler's method to solve the initial value problems with initial conditions $y(k) = 0$ over the interval $[k, 10]$ for $k = 1, 2, 3, 4, 5$. Plot these solutions on the same axes. (Note that the slope of the solution is infinite at $x = k$. The differential equation can be rewritten in the form $dx/dy = y/x$ for these cases. This can be solved in the same way as before – except that we do not know the y-interval over which we must compute the solution.)
9. Consider Newton's Law of cooling as a differential equation model,

$$T'(t) = k(T_E - T(t)),$$

where $TE = 24.5\,°\text{C}$ and $T_0 = 93.5\,°\text{C}$. Suppose $k = 0.25$.

(a) Derive the analytic solution for $T(t)$.
(b) Consider solving the problem numerically using Euler's method to a final time of 30 min and calculating the absolute error at the final time. Solve the problem repeatedly using a sequence of steplengths with $h = 2^{-k}$ and $k = 3, 4, \ldots, 10$, calculating the error for each steplength. What do you see?
(c) Calculate the ratios of successive errors using your results from (b) and explain what you see.

10. A rumor is spread when "information" passes from person to person. One assumption to understand how fast a rumor spreads is; people acquire the rumor by means of public sources, such as the Internet and the rate of acquisition is proportional to the number of people that don't yet know the rumor.

(a) Develop a mathematical model using a differential equation to model the spread of a rumor.

(b) Assume your proportionality constant in the model is 1. Solve the resulting differential equation if the population we are interested in consists of 350,000 people under the following scenarios;

- nobody in the population knows the rumor
- 25,000 initially people know the rumor
- everybody initially knows the rumor

Explain the behavior of the curves.

(c) If the rumor is spread by word of mouth, the rate of acquisition is proportional to the product of those who know the rumor and those who don't. Why does this make sense?

(d) Write a differential equation that models the number of people who have acquired the rumor.

(e) Solve the new problem numerically using the initial conditions that

- 1 person knows the rumor
- that 25,000 people know the rumor

Plot your solutions over time and explain the behavior of the graphs.

11. The above model is not particularly realistic. Develop your own model, using a differential equation, that describes the rate of spreading of a rumor. You may want to revisit the modeling ideas from Chap. 1. Try to solve your problem numerically. What challenges arise during this process. If you are able to get a solution, what are the strengths and weaknesses of your modeling approach?

7.2 Runge–Kutta Methods

The *Runge–Kutta methods*, which we explore next, use additional points between x_k and x_{k+1} which allow the resulting approximation to agree with more terms of the Taylor series. This approach can also be derived by using higher-order numerical integration methods to approximate

$$y(x_{k+1}) - y(x_k) = \int_{x_k}^{x_{k+1}} f(x, y)\, dx$$

The general derivation of Runge–Kutta methods is somewhat complicated so the process is illustrated here with examples of second-order Runge–Kutta methods which are derived from a second-order Taylor series expansion. Higher-order, especially fourth-order, Runge–Kutta methods are commonly used in practice. The basic ideas behind these are similar to those presented. We shall discuss the most commonly used fourth-order Runge–Kutta method in some detail without deriving it in detail.

Consider the second-order Taylor expansion about the point (x_0, y_0) using a steplength h:

$$y(x_1) \approx y_0 + hy_0' + \frac{h^2}{2}y_0''$$

which give an approximate value of

$$y_1 = y_0 + hf(x_0, y_0) + \frac{h^2}{2}y_0'' \tag{7.9}$$

This formula has error of order $O\left(h^3\right)$. However, an approximation for y_0'' is required for this expression to be useful. Since this term is multiplied by h^2, it follows that an approximation to y_0'' which is itself accurate to order $O(h)$ will maintain the overall error in (7.9) at $O\left(h^3\right)$.

For the sake of notation moving forward, denote the slope $y_0' = f(x_0, y_0)$ by k_1. Also let $\alpha \in [0, 1]$ and consider an "Euler step" of length αh. This gives

$$y(x_0 + \alpha h) \approx y_0 + \alpha h k_1$$

and denote by k_2 the slope at the point $(x_0 + \alpha h, y_0 + \alpha h k_1)$ so that $k_2 = f(x_0 + \alpha h, y_0 + \alpha h k_1)$. The simplest divided difference estimate of y_0'' is then given by

$$\begin{aligned}
y_0'' &\approx \frac{y'(x_0 + \alpha h) - y'(x_0)}{\alpha h} \\
&\approx \frac{hf(x_0 + \alpha h, y_0 + \alpha h k_1) - f(x_0, y_0)}{\alpha h} \\
&= \frac{k_2 - k_1}{\alpha h}
\end{aligned} \tag{7.10}$$

Substituting this approximation into (7.9), gives

$$\begin{aligned}
y_1 &= y_0 + hk_1 + \frac{h^2}{2}\frac{k_2 - k_1}{\alpha h} \\
&= y_0 + h\left[k_1\left(1 - \frac{1}{2\alpha}\right) + \frac{k_2}{2\alpha}\right]
\end{aligned} \tag{7.11}$$

which is the general form of the Runge–Kutta second-order formulas. The corresponding formula for the nth step in the solution process is

$$y_{n+1} = y_n + h\left[k_1\left(1 - \frac{1}{2\alpha}\right) + \frac{k_2}{2\alpha}\right] \tag{7.12}$$

where now

$$k_1 = f(x_n, y_n)$$
$$k_2 = f(x_n + \alpha h, y_n + \alpha h k_1) \tag{7.13}$$

Because the approximation (7.10) has error of order $O(h)$, it follows that the local truncation error in (7.11) is $O(h^3)$ as desired.

For implementation, a user has choices for the values of α. There are three common choices, $\alpha = 1/2$, 1, or 2/3, which are used in (7.12) to give different numerical schemes.

$\alpha = 1/2$: **Corrected Euler, or Midpoint, method**

$$y_{n+1} = y_n + h k_2 \tag{7.14}$$

where $k_2 = f(x_n + h/2, y_n + h k_1/2)$. This method is simply the application of the midpoint rule for integration of $f(x, y)$ over the current step with the (approximate) value at the midpoint being obtained by a preliminary (half-) Euler step.

$\alpha = 1$: **Modified Euler method**

$$y_{n+1} = y_n + \frac{h}{2}(k_1 + k_2) \tag{7.15}$$

with $k_2 = f(x_n + h, y_n + h k_1)$. This method results from using the trapezoid rule to estimate the integral where a preliminary (full) Euler step is taken to obtain the (approximate) value at x_{n+1}.

$\alpha = 2/3$: **Heun's method**

$$y_{n+1} = y_n + \frac{h}{4}(k_1 + 3k_2) \tag{7.16}$$

with $k_2 = f(x_n + 2h/3, y_n + 2h k_1/3)$. Heun's method is different in that it cannot be derived from a numerical integration approach.

The first two of these are illustrated in Figs. 7.4 and 7.5. In each of those figures, the second curve represents the solution to the differential equation passing through the appropriate point. For the corrected Euler this "midpoint" slope k_2 is then applied for the full step from x_0 to x_1. For the modified Euler method, the slope used is the average of the slope k_1 at x_0 and the (estimated) slope at x_1, k_2. One can see that both these methods result in significantly improved estimates of the average slope of the true solution over $[x_0, x_1]$ than Euler's method which just uses k_1.

As mentioned above, the local truncation error for all these "second-order" Runge–Kutta methods is $O(h^3)$ but they are called "second-order" methods. The reason for this is that by the time we have computed our approximate value of $y(b) = y(a + Nh)$, we have committed N of these local truncation errors (just like we saw with Euler's method). The effect is that the global truncation error is of order $O(Nh^3)$ which is $O(h^2)$ since $Nh = b - a$ is constant.

Fig. 7.4 Corrected Euler (midpoint method)

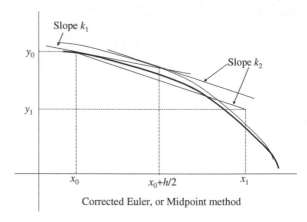

Corrected Euler, or Midpoint method

Fig. 7.5 Modified Euler

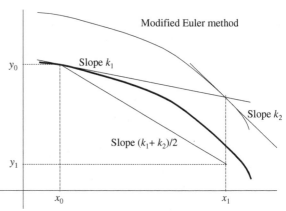

It follows that reducing the steplength by a factor of 1/2 should now result in approximately a 75% improvement in the final answer. That is the error should be reduced by a factor of (about) 1/4. We shall verify this with examples.

Example 2 Use the second-order Runge–Kutta methods (7.14)–(7.16) with $N = 2$ to solve the initial value problem $y' = (6x^2 - 1)y; \ y(0) = 1$ on $[0, 1]$.

For the Corrected Euler (midpoint) (7.14) method we obtain the results below. The true solution ($y = e^{2x^3 - x}$) is tabulated for comparison.

k	Corrected Euler	True solution value
0	$k_1 = f(0, 1) = -1$ $k_2 = f(1/4, 3/4) = -0.4688$ $y_1 = 1 + (1/2)(3/16) = 0.7656$	$\exp(2*(1/2)^3 - 1/2) = 0.7788$
1	$k_1 = f(1/2, 0.7788) = 0.3828$ $k_2 = f(3/4, 0.8613) = 2.0457$ $y_2 = 0.7788 + (1/2)(2.0457) = 1.7885$	$\exp(1) = 2.7183$

The other two methods yield the results tabulated here in somewhat less detail.

k	Modified Euler	Heun's method
0	$k_1 = -1$ $k_2 = 0.25$ $y_1 = 0.8125$	$k_1 = -1$ $k_2 = -0.2222$ $y_1 = 0.7917$
1	$k_1 = 0.4063$ $k_2 = 5.0781$ $y_2 = 2.1836$	$k_1 = 0.3958$ $k_2 = 2.9248$ $y_2 = 1.0379$

All of these methods are easily programmed. (See the Exercises.) The values with only 2 steps may look somewhat pessimistic at first, but compare to the forward Euler methods, which took roughly 8 steps to get to nearly the same accuracy.

Example 3 Apply the three Runge–Kutta methods (7.14)–(7.16) to the same initial value problem with $N = 10$ and $N = 100$. Compare their errors and verify that the global truncation error is $O(h^2)$.

The intermediate values for using $N = 10$ are shown below.

x	Corrected	Modified	Heun
0	1.0000	1.0000	1.0000
0.1000	0.9064	0.9077	0.9069
0.2000	0.8317	0.8338	0.8324
0.3000	0.7817	0.7844	0.7826
0.4000	0.7615	0.7648	0.7626
0.5000	0.7778	0.7824	0.7794
0.6000	0.8428	0.8496	0.8451
0.7000	0.9796	0.9908	0.9834
0.8000	1.2349	1.2549	1.2415
0.9000	1.7052	1.7441	1.7180
1.0000	2.6033	2.6850	2.6301

For this particular example, the modified Euler method appears to be performing somewhat better than the other two. The errors at $x = 1$ are

Corrected Modified Heun
0.1150   0.0333 0.0882

For comparison the error in Euler's method for $N = 10$ is 0.9358 meaning second order Runge–Kutta gives a significant improvement.

For $N = 100$, we of course do not tabulate all the results. The errors at $x = 1$ are now

Corrected   Modified   Heun
1.5477×10^{-3} 3.7090×10^{-4} 1.1559×10^{-3}

and for Euler's method the error is 1.3222×10^{-1}.

With the increase in N by a factor of 10 we should expect the error in the first-order Euler's method to be reduced by approximately this same factor. The errors in the second order methods should be reduced by a factor of about 10^2, which is shown in the results.

For the particular equation in Example 3, the modified Euler method (which uses an approximate trapezoid rule to integrate the slope) appears to be the most accurate. This may not always be the case and in for this example, it is likely due to the fact that the function is concave up, so that using slope estimates from further to the right is likely to be beneficial.

One of the most widely used Runge–Kutta methods is the classical fourth-order formula, RK4

$$y_{n+1} = y_n + \frac{h}{6} [k_1 + 2 (k_2 + k_3) + k_4] \tag{7.17}$$

where the various slope estimates are

$$
\begin{aligned}
k_1 &= f (x_n, y_n) \\
k_2 &= f (x_n + h/2, y_n + hk_1/2) \\
k_3 &= f (x_n + h/2, y_n + hk_2/2) \\
k_4 &= f (x_n + h, y_n + hk_3)
\end{aligned}
$$

The derivation of this result, and of the fact that its global truncation error is $O\left(h^4\right)$ are not included here. The basic idea is again that of obtaining approximations which agree with more terms of the Taylor expansion.

This approach can also be viewed (7.17) as an approximate application of Simpson's rule to the integration of the slope over $\left[x_n, x_{n+1}\right]$. Here k_1 is the slope at x_n, k_2 and k_3 are both estimated slopes at the midpoint $x_n + h/2$, and k_4 is then an estimate of this slope at $x_n + h = x_{n+1}$. With this interpretation, (7.17) is equivalent to Simpson's rule where the value at the midpoint is replaced by the average of the two estimates. This gives insight to the truncation error, recall that we already saw that

Simpson's rule has an error of order $O\left(h^4\right)$. Moreover, the second-order Runge–Kutta methods (7.14) and (7.15) have errors of the same order as their corresponding integration rules.

Program   Python code for the classical Runge–Kutta fourth-order method RK4
(7.17).

```python
import numpy as np

def RK4(fcn, a, b, y0, N):
    """
    Solve y' = f(x,y) in 'N' steps using
    fourth-order Runge-Kutta with initial
    condition y[a] = y0.

    """

    h = (b-a) / N
    x = a + np.arange(N+1) * h
    y = np.zeros(x.size)
    y[0] = y0
    for k in range(N):
        k1 = fcn(x[k], y[k])
        k2 = fcn(x[k] + h/2, y[k] + h * k1/2)
        k3 = fcn(x[k] + h/2, y[k] + h * k2/2)
        k4 = fcn(x[k] + h, y[k]  + h * k3)
        y[k + 1] = y[k] + h * (k1 + 2 * (k2 + k3)
                            + k4) / 6

    return x, y
```

Example 4 Apply RK4 to our usual example with both $N = 10$ and $N = 100$.

Using $N = 10$ produces the results tabulated.

x	RK4 solution	True solution
0	1.0000e+00	1.0000e+00
1.0000e−01	9.0665e−01	9.2312e−01
2.0000e−01	8.3194e−01	8.8692e−01
3.0000e−01	7.8192e−01	8.8692e−01
4.0000e−01	7.6185e−01	9.2312e−01
5.0000e−01	7.7880e−01	1.0000e+00
6.0000e−01	8.4535e−01	1.1275e+00
7.0000e−01	9.8609e−01	1.3231e+00
8.0000e−01	1.2511e+00	1.6161e+00
9.0000e−01	1.7471e+00	2.0544e+00
1.0000e+00	2.7179e+00	2.7183e+00

The results show much better accuracy than the previous approaches and the error in $y(1)$ is 4.2884×10^{-4}

With $N = 100$, this error is reduced to only 5.1888×10^{-8} which reflects the reduction by a factor of 10^4 which we should anticipate for a fourth-order method when the number of steps is increased by a factor of 10.

Higher-order Runge–Kutta methods can give high-accuracy answers with small numbers of steps for a wide range of differential equations. There are however situations when even these methods are insufficient. The so-called *stiff* differential equations are one such case.

Exercises

Exercises 1–5, and 8 concern the differential equation

$$y' = x/y$$

1. For the initial condition $y(0) = 3$, use the corrected Euler's method with steps $h = 1, 1/2$ and $1/4$ to approximate $y(1)$.
2. Repeat Exercise 1 for the modified Euler and Heun's methods.
3. Write a script to implement the three second-order Runge–Kutta methods we have discussed. Use them to solve the initial value problem of Exercises 1 and 2 over $[0, 4]$ with $N = 10, 20, 50, 100$ steps.
4. Tabulate the errors in the approximate values of $y(4)$ in Exercise 3. Verify that their errors appear to be $O(h^2)$. Which of the methods appears best for this problem?
5. Repeat Exercise 3 for different initial conditions: $y(0) = 1, 2, 3, 4, 5$ and plot the solutions for $N = 100$ on the same axes.
6. Show that if $f(x, y)$ is a function of x alone, then the modified Euler method is just the trapezoid rule with step h.
7. Use the classical Runge–Kutta RK4 method with steplengths $h = 10^{-k}$ for $k = 1, 2, 3$ to solve the initial value problem $y' = x + y^2$ with $y(0) = 0$ on $[0, 1]$. Tabulate the results for $x = 0, 0.1, 0.2, \ldots, 1$ and graph the solutions.

8. Repeat Exercise 7 for the initial value problem of Exercise 1. Verify that the errors are of order $O\left(h^4\right)$.

9. Show that if $f(x, y)$ is a function of x alone, then RK4 is just Simpson's rule with step $h/2$.

10. Solve the differential equation $y' = -x \tan y$ with $y(0) = \pi/6$ for $x \in [0, 1]$. Compare your solution at the points between 0 and 1 using increments of 0.1 with the values obtained using RK4 with $h = 0.1$ and $h = 0.05$.

11. Repeat Exercise 9 from Sect. 7.1 using the methods from this section. How do your errors compare?

12. Star Trek fans may remember when the cute, fuzzy Tribbles quickly became trouble on the starship Enterprise. One Tribble was trapped in the ship and a single Tribble actually produces in a litter of 10 Tribbles every 12 h. Mr. Spock claimed that after 3 days, there were already 1,771,561 Tribbles.

(a) Was Spock correct?
(b) Use mathematical modeling to represent the way the Tribble population changes over time as a differential equation.
(c) Solve the differential equation numerically and provide a plot of the Tribble population for a week.
(d) When was the Tribble population 500,000? will the Tribble population reach 4,000,000?
(e) Experiment with your model and incorporate a removal mechanism. At which rate would the Tribbles need to be removed to have them all off the ship in a day?

7.3 Multistep Methods

The methods considered next are like the Runge–Kutta methods in that they use more than one point to compute the estimated value of y_{k+1}. They differ in that all the information is taken from tabulated points.

The simplest example is based on a second-order Taylor expansion

$$y_{n+1} = y_n + hy'_n + \frac{h^2}{2}y''_n = y_n + hf(x_n, y_n) + \frac{h^2}{2}y''_n. \tag{7.18}$$

Using a backward difference approximation to the second derivative gives

$$y''_n \approx \frac{y'(x_n) - y'(x_n - h)}{h} \approx \frac{f(x_n, y_n) - f(x_{n-1}, y_{n-1})}{h}$$

and substituting this into (7.18) we obtain the *two-step formula*

$$y_{n+1} = y_n + hf(x_n, y_n) + \frac{h^2 \left[f(x_n, y_n) - f(x_{n-1}, y_{n-1}) \right]}{2h}$$

$$= y_n + \frac{h}{2}(3f_n - f_{n-1}) \tag{7.19}$$

where we have abbreviated $f(x_n, y_n)$ to just f_n. This particular formula is the two-step *Adams–Bashforth* method. The formula (7.19) agrees with the Taylor expansion as far as the second-order term and so has local truncation error of order $O(h^3)$, and consequently, the global truncation error is $O(h^2)$.

In this text, we cover only the class of multistep methods known as Adams methods, which includes the Adams–Bashforth methods as one of their two important subclasses. The other is discussed later. A general N-step Adams method uses values from N previous points in order to estimate y_{n+1}. The general formula is therefore

$$y_{n+1} = y_n + h \sum_{k=0}^{N} \beta_k f_{n+1-k} \tag{7.20}$$

The coefficients β_k are chosen to give the maximum order of agreement with the Taylor expansion. Like the Runge–Kutta methods, these Adams methods can be viewed as the application of numerical integration rules but differ in that nodes are incorporated from outside the immediate interval. The cases where $\beta_k = 0$ are the general N-step Adams–Bashforth methods which provide *explicit* formulas for y_{n+1}.

An important observation in (7.20) is that since the coefficient β_0 multiplies the slope $f_{n+1} = f(x_{n+1}, y_{n+1})$, then if $\beta_0 \neq 0$, the right-hand side of (7.20) depends on the very quantity y_{n+1} which we are trying to approximate. This gives an implicit formula for y_{n+1} which could be solved iteratively (perhaps with a nonlinear solver that we covered earlier). Adams methods with $\beta_0 \neq 0$ are known as *Adams–Moulton* methods.

Let's derive the Adams–Moulton method which agrees with the Taylor expansion up to third order. We start with

$$y(x_{n+1}) \approx y_n + hy'_n + \frac{h^2}{2}y''_n + \frac{h^3}{6}y'''_n \tag{7.21}$$

and can then use difference approximations to the higher derivatives. Since the Adams–Moulton formulas are implicit, we can use the (better) symmetric approximations to the derivatives:

$$y''_n \approx \frac{y'_{n+1} - y'_{n-1}}{2h} = \frac{f_{n+1} - f_{n-1}}{2h}$$

and

$$y'''_n \approx \frac{y'_{n+1} - 2y'_n + y'_{n-1}}{h^2} = \frac{f_{n+1} - 2f_n + f_{n-1}}{h^2}$$

(Recall from the Numerical Calculus section) Substituting these into (7.21) gives

$$y_{n+1} = y_n + hf_n + \frac{h^2}{2}\frac{f_{n+1} - f_{n-1}}{2h} + \frac{h^3}{6}\frac{f_{n+1} - 2f_n + f_{n-1}}{h^2}$$

$$= y_n + \frac{h}{12}\left(5f_{n+1} + 8f_n - f_{n-1}\right) \tag{7.22}$$

which is the two-step Adams–Moulton formula.

The two-step Adams–Moulton formula (7.22) agrees with the third-order Taylor expansion and so has local truncation error $O\left(h^4\right)$ and a corresponding global truncation error of order $O\left(h^3\right)$. This illustrates the potential benefit of the Adams–Moulton methods. The N-step Adams–Moulton method has error of order one greater than the corresponding N-step Adams–Bashforth method.

The second-order Adams–Moulton formula is the one-step method which corresponds to the implicit trapezoid rule

$$y_{n+1} = y_n + \frac{h}{2}\left(f_{n+1} + f_n\right) \tag{7.23}$$

When it comes to implementation, a user has choices about how to get multi-step methods started. To use the two-step Adams–Bashforth formula (7.19) to obtain y_1 appears to require knowledge of y_{-1} which would not be available. For higher-order methods using more steps, even more points are needed. The usual solution to this problem is that a small number of steps of a Runge–Kutta method of the desired order could generate enough values to allow the Adams methods to proceed.

As an example, the modified Euler method (a second-order Runge–Kutta method) could generate y_1 after which the second-order (two-step) Adams–Bashforth method would be used to generate $y_2, y_3, \ldots$. For a fourth order Adams method, one could use RK4 to generate y_1, y_2, y_3 which is enough points to continue with the Adams method.

In terms of computational efficiency, once the Adams methods are started, subsequent points are generated without the need for any of the intermediate points required by the Runge–Kutta methods. For example, the second-order Adams–Bashforth method would use only about half the computational effort of a second-order Runge–Kutta method with the same basic steplength. this is a general advantage of Adams methods.

Example 5 Use the two-step Adams–Bashforth method to solve the initial value problem $y' = (6x^2 - 1)y$; $y(0) = 1$ on [0, 1] with $h = 1/4$.

A second-order Runge–Kutta method is first used to get y_1. The modified Euler method yields

$$y_1 = 0.8164$$

The Adams–Bashforth method can now be used to generate the remaining values as follows:

x	y	f
0	1	-1
1/4	0.8164	-0.5103
1/2	$0.8164 + (1/8)\,(3\,(-0.5103) - (-1)) = 0.7501$	0.3750
3/4	$0.7501 + (1/8)\,(3\,(0.3750) - (-0.5103)) = 0.9545$	2.2669
1	$0.9545 + (1/8)\,(3\,(2.2669) - 0.3750) = 1.7577$	8.7884

Example 6 Use the Adams–Bashforth two-step method to solve the same initial value problem as in Example 5 using $N = 10$ steps.

This is easy to program. The Python function used was as follows.

```python
import numpy as np

def ab2(fcn, a, b, y0, N):
    """
    Solve y' = f(x,y) on [a,b] in 'N' steps using
    two-step Adams-Bashford, with initial
    condition y[a] = y0.

    """

    h = (b - a) / N
    x = a + np.arange(N+1) * h
    y = np.zeros(x.size)
    f = np.zeros_like(y)
    y[0] = y0
    f[0] = fcn(x[0], y[0])
    # Use modified Euler for first step.
    k1 = f[0]
    k2 = fcn(x[0] + h, y[0] + h * k1)
    y[1] = y[0] + h * (k1 + k2) / 2
    f[1] = fcn(x[1], y[1])
    # Use two-step formula for remaining points
    for k in range(1,N):
        y[k + 1] = y[k] + h * (3 * f[k] - f[k-1])/2
        f[k + 1] = fcn(x[k + 1], y[k + 1])

    return x, y
```

Table 7.1 Results for Adams methods for the initial value problem of Example 6

x	AB2, $N = 10$	ModEuler, $N = 10$	AB2, $N = 20$	ABM23, $N = 10$
0	1.0000	1.0000	1.0000	1.0000
0.1000	0.9077	0.9077	0.9063	0.9077
0.2000	0.8338	0.8297	0.8310	0.8329
0.3000	0.7844	0.7778	0.7805	0.7828
0.4000	0.7648	0.7557	0.7599	0.7628
0.5000	0.7824	0.7690	0.7757	0.7800
0.6000	0.8496	0.8282	0.8400	0.8468
0.7000	0.9908	0.9531	0.9756	0.9878
0.8000	1.2549	1.1824	1.2288	1.2525
0.9000	1.7441	1.5936	1.6956	1.7455
1.0000	2.6850	2.3484	2.5886	2.7022

The choice of the modified Euler method to generate y_1 (or y[1] in the Python code) is somewhat arbitrary, any of the second-order Runge–Kutta methods could have been used.

The results are included in the second column of Table 7.1. For comparison, the third column of Table 7.1 shows the results obtained using the modified Euler method (Example 3) with the same number of steps. These two methods have global truncation errors of the same order $O\left(h^2\right)$. The explicit Adams–Bashforth method appears much inferior. *But* we must note that each step of AB2 requires only one new evaluation of the right-hand side, whereas the modified Euler method (or any second-order Runge–Kutta method) requires two such evaluations per step.

It would therefore be fairer to compare the two-step Adams–Bashforth method using $N = 20$ with the modified Euler method using $N = 10$. These results form the third set of output. As expected the error is reduced by (approximately) a factor of 1/4 compared to $N = 10$. Although the error is still greater than that for the modified Euler method, it is now comparable with those for the other second-order Runge–Kutta methods in Example 3.

Until now we haven't discussed the role of Adams–Moulton methods. Since they are implicit, they require more computational effort that the Adams–Bashforth methods. A practical approach is to use them in conjunction with Adams–Bashforth methods as *predictor-corrector* pairs. The basic idea is that an explicit method is used to estimate y_{n+1}, after which this estimate can be used in the right-hand side of the Adams–Moulton (implicit) formula to obtain an improved estimate. Thus the Adams–Bashforth method is used to *predict* y_{n+1} and the Adams–Moulton method is then use to *correct* this estimate.

Example 7 Use an Adams predictor-corrector method to solve the usual example problem. Specifically, use the two two-step methods (7.19) and (7.22) with $N = 10$ steps.

The code for the two-step Adams–Bashforth method is readily modified to add the corrector step. The basic loop becomes

```
for  k  in  range (1 ,N) :
    y [k  +  1]  =  y [k]  +  h  *  (3  *  f [k]  −  f [k−1])  /  2
    f [k  +  1]  =  fcn (x [k  +  1],  y [k  +  1])
    y [k  +  1]  =  y [k]  +  h  *  (5  *  f [k  +  1]  +  8  *  f [k]
                          −  f [k−1])  /  12
    f [k  +  1]  =  fcn (x [k  +  1],  y [k  +  1])
```

The results are shown as the final column of Table 7.1. The predictor-corrector method provides smaller error than the modified Euler method using the same number of points for this example. This is a fair comparison since both methods entail two new evaluations of the slope per step.

To re-emphasize the connection between differential equations and numerical integration formulas, let's revisit the derivation of the coefficients for the Adams methods. If $f(x, y)$ has no explicit dependence on y then it follows that

$$y_{n+1} - y_n = \int_{x_n}^{x_{n+1}} f(x)\,dx = h \int_0^1 f(x_n + ht)\,dt.$$

So it is possible to find values of the coefficients in (7.20) so that the corresponding quadrature formula has maximal degree of precision. (Note here that the integration formula includes some nodes from outside the range of integration corresponding to $t = -1, -2, \ldots$.)

For a three-step Adams–Bashforth formula we therefore seek $\beta_1, \beta_2, \beta_3$ so that the quadrature formula

$$\int_0^1 F(t)\,dt \approx \beta_1 F(0) + \beta_2 F(-1) + \beta_3 F(-2)$$

is exact for all quadratic functions F. Imposing this condition for $F(t) = 1, t, t^2$ in turn we get the equations

$$
\begin{aligned}
F(t) = 1 &: \beta_1 + \beta_2 + \beta_3 = 1 \\
F(t) = t &: -\beta_1 - 2\beta_2 = 1/2 \\
F(t) = t^2 &: \beta_2 + 4\beta_3 = 1/3
\end{aligned}
$$

Adding the last two of these yields $\beta_3 = 5/12$, and thence we obtain $\beta_2 = -4/3$, $\beta_1 = 23/12$. Therefore the three-step Adams–Bashforth formula is

$$y_{n+1} = y_n + \frac{h(23 f_n - 16 f_{n-1} + 5 f_{n-2})}{12} \tag{7.24}$$

(Note that the coefficients must sum to one, providing an easy check.)

The three-step formula (7.24) has fourth-order local truncation error so that its global truncation error is of order $O\left(h^3\right)$.

The corresponding three-step (fourth-order) Adams–Moulton formula is based on the quadrature formula

$$\int_0^1 F\left(t\right) dt \approx \beta_0 F\left(1\right) + \beta_1 F\left(0\right) + \beta_2 F\left(-1\right) + \beta_3 F\left(-2\right)$$

Using a similar approach to that above gives

$$y_{n+1} = y_n + \frac{h\left(9 f_{n+1} + 19 f_n - 5 f_{n-1} + f_{n-2}\right)}{24} \tag{7.25}$$

A modification of the code used in Examples 6 and 7 could be used to implement a three-step predictor corrector method. The corrector formula has $O\left(h^4\right)$ truncation error and so we might compare its performance with the fourth-order Runge–Kutta formula.

Exercises

Exercises 1–6 and 10–12 concern the initial value problem $y' = x/y$, $y\left(0\right) = 3$

1. Apply the two-step Adams–Bashforth method AB2 with $N = 4$ on $[0, 1]$. Use the modified Euler method for the first step.
2. Compare AB2 using $N = 20$ steps with the modified and corrected Euler methods using $N = 10$ steps on $[0, 4]$. (The computational effort is comparable in all these.)
3. Repeat Exercise 2 for $N = 40, 100, 200$ for AB2 and $N = 20, 50, 100$ for the Runge–Kutta methods. Verify that the truncation error of AB2 is second-order.
4. Use the predictor-corrector method with AB2 as predictor and the second-order Adams–Moulton formula AM2 (implicit trapezoid rule) as corrector. Take $N = 10$ and compare the results with those of Exercise 2.
5. Repeat Exercise 4 using the third order Adams–Moulton formula as corrector.
6. Use $N = 20, 50, 100$ to compare the methods of Exercises 4 and 5 with the Runge–Kutta methods of Exercise 3. What are your estimates of the order of the truncation errors for the two predictor-corrector methods?
7. Derive the second-order Adams–Bashforth method as a quadrature formula. That is find the quadrature rule

$$\int_0^1 F\left(t\right) dt \approx \beta_1 F\left(0\right) + \beta_2 F\left(-1\right)$$

   which is exact for $F\left(t\right) = 1, t$.
8. Derive the fourth-order Adams–Moulton formula AM4. (This will use nodes $1, 0, -1, -2$ and be exact for cubic polynomials. See Eq. (7.25).)

9. Derive the fourth-order Adams–Bashforth formula. (This will use nodes $0, -1$, $-2, -3$ and be exact for cubic polynomials.)

10. Apply the third and fourth order Adams–Bashforth methods AB3 and AB4 to the standard problem using $N = 20$ steps on $[0, 4]$. Compare the results with those of the fourth-order Runge–Kutta method RK4 with $N = 10$.

11. Repeat Exercise 10, doubling the numbers of steps used twice ($N = 40$ and $N = 80$ for the Adams–Bashforth methods). What are your estimates of the orders of their truncation errors?

12. Implement the Adams predictor-corrector method. Compare the performance to using the fourth-order Runge–Kutta method the usual example problem $y' = 3x^2 y$; $y(0) = 1$ on $[0, 1]$. Use 10 and then 20 steps and explain what you see. Calculate the ratio of successive errors and explain what you see.

13. Use the predictor-corrector methods with AB3 and AB4 as predictors and AM4 as corrector. Use $N = 20, 40, 80$. Compare the results with those for RK4 using $N = 10, 20, 40$.

7.4 Systems of Differential Equations

We now focus on systems of differential equations, which are used to model interdependent scenarios. Models that couple quantities that are interacting and changing over time are often approached mathematically in this way. The first model we discussed in this text, that of the spread of a disease, is an example where three populations interact and change; susceptible individuals, infected individuals, and recovered individuals (often called the SIR model). Given an initial population containing a certain number of sick individuals, one can predict how those three populations are changing in time under certain assumptions about the rates that the disease spreads and people recover. Another common model that uses systems of differential equations is the predator-prey model from population dynamics. We provide an example of this shortly.

Consider the problem of solving a system of first-order initial value problems such as

$$\mathbf{y}' = \mathbf{f}(x, \mathbf{y}), \quad \mathbf{y}(x_0) = \mathbf{y}_0, \tag{7.26}$$

which is written in vector-form. Here $\mathbf{y}$ is an n-dimensional vector function of x, so that $\mathbf{y} : \mathfrak{R} \to \mathfrak{R}^n$ and $\mathbf{f}$ is an n-dimensional vector-valued function of $n + 1$ variables, $\mathbf{f} : \mathfrak{R}^{n+1} \to \mathfrak{R}^n$. Note that we are making no distinction here between the *function* and its *values*. The meaning should hopefully be clear from the context. It may help to view the component form (7.26), given by

$$
\begin{aligned}
y_1' &= f_1(x, y_1, y_2, \ldots, y_n), & y_1(x_0) &= y_{1,0} \\
y_2' &= f_2(x, y_1, y_2, \ldots, y_n), & y_2(x_0) &= y_{2,0} \\
&\cdots \quad \cdots \quad \cdots \\
y_n' &= f_n(x, y_1, y_2, \ldots, y_n), & y_n(x_0) &= y_{n,0}
\end{aligned}
\tag{7.27}
$$

The techniques available for initial value problems of this sort are essentially the same as those for single differential equations that we have already discussed. The primary difference is that, at each step, a *vector* step must be taken.

Since the classical fourth-order Runge–Kutta method RK4 is both straightforward to program and efficient, let's concentrate on its use for solving systems. The python code below is almost identical to that in Sect. 7.2. The only difference is the use of vector quantities for the initial conditions and in the main loop.

Program   RK4 for an initial value system.

```python
def RK4sys(fcn, a, b, y0, N):
    """
    Solve y' = f(x,y) on the interval [a,b] in
    'N' steps for a system of differential
    equations using fourth-order Runge-Kutta with
    initial condition y[a] = y0 where 'y0' is a
    vector.

    """

    h = (b - a) / N
    x = a + np.arange(N+1) * h
    y = np.zeros((x.size, y0.size))
    y[0, :] = y0
    for k in range(N):
        k1 = fcn(x[k], y[k, :])
        k2 = fcn(x[k] + h/2, y[k, :] + h * k1 / 2)
        k3 = fcn(x[k] + h/2, y[k, :] + h * k2 / 2)
        k4 = fcn(x[k] + h, y[k, :] + h * k3)
        y[k+1, :] = y[k, :] + h * (k1 + 2
                    * (k2 + k3) + k4) / 6

    return x, y
```

Example 8 Predator-Prey model. Let $R(t)$, $F(t)$ represent the populations of a rabbits and foxes at time t in a closed ecosystem. We assume that they behave like a simple food chain in that the rabbits feed on grass and the foxes feed on the rabbits. A simple predator-prey model for this system can be used to predict the sizes of the two populations. It has the form

$$R'(t) = \alpha R(t) - \beta R(t) F(t)$$
$$F'(t) = -\gamma F(t) + \delta F(t) R(t)$$

Here α represents the rate at which the rabbit population would grow in the absence of foxes, γ is the rate at which the fox population would decline in the absence of the rabbits. The two nonlinear terms represent the interaction of these two populations. All the coefficients are assumed positive, with the attached signs implying the "direction" of their influence. With t measured in months, find the populations of rabbits and foxes after 4 months given initial populations

$$R(0) = 500, \quad F(0) = 200$$

with $\alpha = 0.7, \gamma = 0.2, \beta = 0.005, \delta = 0.001$. (The first two of these imply that the unchecked rabbit population would more than double in one month, and the unchallenged fox population declines about 20% per month.)

The right-hand side of this system is represented by the function

```
def predprey(t, y):
    r = np.empty(2)
    r[0] = 0.7 * y[0] - 0.005 * y[0] * y[1]
    r[1] = -0.2 * y[1] + 0.001 * y[0] * y[1]

    return r
```

where y[0] represents the rabbit population R, and y[1] represents the foxes F.

The solution is then computed by the commands

```
>>> RF0 = np.array([500, 200])
>>> sol_x, sol_y = RK4sys(predprey, 0, 4, RF0, 100)
```

A meaningful visualization of the solution is obtained by plotting the rabbit population against the fox population at the corresponding time:

```
>>> plt.plot(sol_y[:,1], sol_y[:,0], 'k-')
```

The resulting plot is shown in Fig. 7.6. It is evident that the rabbit population declines steadily while the foxes multiply until their population peaks at about 260. At this time the foxes appear to have "won" and both populations decrease.

Fig. 7.6 Solution of the
predator-prey model

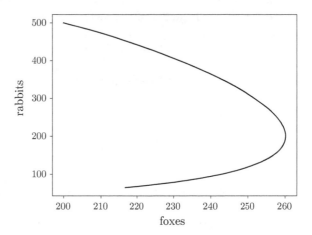

Another important class of differential equations, higher-order equations, can actually be rearranged as systems of first-order equations and then can then be solved using the same techniques. Consider the principle for a second-order initial value problem which has the general form

$$y'' = f\left(x, y, y'\right); \quad y\left(x_0\right) = y_0, \ y'\left(x_0\right) = y_0'$$ \hfill (7.28)

We can define two new variables (functions) by

$$u_1 = y, \quad u_2 = y'$$

Then, by definition, it follows that

$$u_1' = u_2$$

and the differential equation (7.28) can be rewritten as

$$u_2' = f\left(x, u_1, u_2\right)$$

The initial value problem therefore becomes the first-order system

$$\begin{aligned} u_1' &= u_2 & u_1\left(x_0\right) &= y_0 \\ u_2' &= f\left(x, u_1, u_2\right) & u_2\left(x_0\right) &= y_0' \end{aligned}$$

which could now be solved in just the same way as was used for the predator-prey model in Example 8.

Example 9 Solve the initial value problem

$$y'' = -y; \quad y(0) = 1, y'(0) = 1$$

on the interval $[0, 3]$ using 100 steps of RK4.

First we rearrange the equation as a system:

$$\begin{aligned} u_1' &= u_2 & u_1(0) &= 1 \\ u_2' &= -u_1 & u_2(0) &= 1 \end{aligned}$$

This system can now be solved using the same approach as above. The true solution to this equation is $y = sin(x) + cos(x)$. Simply plotting the approximation and the true solution together would show no apparent differences between the two because the error is significantly small. Indeed, the error in the solution at the first step is roughly 10^{-10} and after 100 steps, although it has grown due the accumulation of error at each step, it is only about 10^{-8}.

Note that the examples used so far have been *initial value* problems. For higher-order differential equations it is common to be given boundary conditions at two distinct points. A technique for solving boundary value problems is the subject of the next section.

Exercises

1. Solve the initial value system $y_1' = x + y_1 - 2y_2$, $y_2' = 2y_1 + y_2$ with $\mathbf{y}(0) = (0, 0)$ over the interval $[0, 4]$ using 200 steps of the fourth-order Runge–Kutta method. Plot the functions y_1, y_2 on the same axes.
2. Solve the same predator-prey equations as in Example 8 over longer time spans $[0, 10]$ and $[0, 20]$ using $h = 0.1$ in both cases.
3. Vary the starting conditions for the predator-prey problem of Exercise 2. Fix $B(0) = 500$ and use $A(0) = 100, 200, 300$ over the intervals $[0, 20]$, $[0, 20]$ and $[0, 30]$ respectively. Plot the solutions on the same axes.
4. As mentioned in the beginning of this section, the spread of a disease (often called an SIR model) can be modeled with a system of differential equations. Investigate these ideas and choose a disease to study. Use the ideas in this section to model how the populations are changing over some time span. Consider a range of different initial conditions and model parameters and explain what you see.
5. Solve the Bessel equation $x^2 y'' + xy' + (x^2 - 1) y = 0$ with initial conditions $y(1) = 1$, $y'(1) = 0$ over the interval $[1, 15]$ using $N = 280$ steps. Plot the computed solution.
6. Solve the second-order nonlinear equation $y'' = 4xy' + 2(1 - 2x^2)y$ over the interval $[0, 2]$ using $N = 100$ steps with the intitial conditions $y(0) = 0$, $y'(0) = 1$. Show that the true solution is $y = xe^{x^2}$ and plot the error. Explain what you see.

7. Solve the second-order nonlinear equation $y'' = 2yy'$ over the interval $[0, 1.5]$ using $N = 150$ steps for the following initial conditions:
   (a) $(y(0), y'(0)) = (0, 1)$ (b) $(y(0), y'(0)) = (0, 2)$ (c) $(y(0), y'(0)) = (1, 1)$ (d) $(y(0), y'(0)) = (1, -1)$.
   Plot all the solutions on the same axes.

8. A projectile is fired with initial speed 1000 m/s at angle α^0 from the horizontal. A good model for the path uses an air-resistance force proportional to the square of the speed. This leads to the equations

$$x'' + cx'\sqrt{x'^2 + y'^2} = 0, \quad y'' + cy'\sqrt{x'^2 + y'^2} = -g$$

   Take the constant $c = 0.005$. The initial conditions are $x(0) = y(0) = 0$ and $(x'(0), y'(0)) = 1500\,(\cos\alpha, \sin\alpha)$. Convert this system of equations to a set of four first-order equations and solve them over the time-interval $[0, 20]$ for a range of launch angles between 10 and 80. Plot the trajectories on the same set of axes.

9. The motion of a object attached to a vertical spring can be modeled with second order differential equations under a variety of different physical scenarios and assumptions (often called a spring-mass system). Investigate these ideas and choose a model to study. Use the ideas in this section to generate a plot the motion of the spring. Consider a range of different initial conditions and model parameters and explain what you see.

10. The motion of a pendulum can be modeled with second order differential equations under a variety of different physical scenarios and assumptions. Investigate these ideas and choose a model to study. Use the ideas in this section to generate a plot the path of the pendulum. Consider a range of different initial conditions and model parameters and explain what you see.

11. Imagine using mathematics to understand if we would survive a zombie outbreak. See the article When *Zombies Attack: Mathematical Modeling of an Outbreak of Zombie Infection*, P. Munz et al., In J.M. Infectious Disease Modelling Research Progress, Tchuenche and C. Chiyaka, editors, 2009 for example.
    Consider the basic model which has three components; susceptible humans (s), zombies (z), and removed individuals (r) (i.e. truly dead people or 'dead' zombies). The coupled system of first order differential equations is given by:

$$s' = \Pi - \beta sz - \delta s$$
$$z' = \beta sz - \gamma r - \alpha sz$$
$$r' = \delta s + \alpha sz - \gamma r$$

Here, susceptible people can die via natural causes at a death rate δ. Here they are taken out of the human population and added to the 'removed' population. Humans in the removed class may become zombies if they are resurrected via the parameter γ. Humans can become zombies if they encounter a zombie via the interaction parameter β. Zombies leaving the model and entering the removed class because a human won a battle with a zombie (by chopping the head off) is accounted for with the parameter α. Π is a birth rate of humans, but we will assume that this attack is happening quickly enough that this can be set to zero.

(a) What are some assumptions that were made to arrive at this model?
(b) Solve this system using RK4 for systems to predict these populations after 5 days. Assume an initial human population of 500, 1 zombie, and 0 removed people. For model parameters, use $\Pi = 0, \alpha = 0.005, \beta = 0.0095, \gamma = 0.0001, \delta = 0.0001$.
(c) Provide a plot of the zombies and the humans over time.
(d) Play with the model parameters and provide some illustrative plots. Are we doomed?

7.5   Boundary Value Problems: Shooting Methods

To motivate the need for the numerical technique in this section, we start with an example and then generalize the solution approach. We consider using a model for projectile motion to determine a launch angle and hit a specific target. This example also demonstrates how several of the ideas in this text can be combined to propose a solution to a real-world problem.

Example 10 Find the angle α at which a projectile must be launched from the origin with an initial speed of 500 m/s to hit a target with coordinates (1000, 150) in meters. Represent this scenario mathematically and outline an approach to solve this problem numerically.

One approach to modeling this while including air resistance has the drag force proportional to the square of the speed. The vector differential equation is then of the form

$$\mathbf{r}'' + cv\mathbf{r}' = -g\mathbf{j}$$

where $v = |\mathbf{r}'|$ is the speed of the projectile. Here c is the constant of proportionality which will depend on many physical parameters including the density and viscosity of the medium (in this example, air), the size, mass, cross-section, and aerodynamic properties of the projectile. This is subject to initial conditions,

$$\mathbf{r}(0) = (0, 0), \mathbf{r}'(0) = (500 \cos \alpha, 50 \sin \alpha).$$

The resulting nonlinear equation cannot be solved exactly and we require numerical methods. Using ideas from the previous section, this model can convert this higher-order system to a system of linear differential equations. Writing the vector differential equation in component form we have

$$x'' + c\sqrt{(x')^2 + (y')^2}x' = 0$$

$$y'' + c\sqrt{(x')^2 + (y')^2}x' = -g$$

Each of these is a second order equation and can be converted to a system of first order equations. Let's define

$$u_1 = x, u_2 = x', u_3 = y, u_4 = y'$$

and then the final system becomes

$$u_1' = u_2$$
$$u_2' = -cu_2\sqrt{u_2^2 + u_4^2}$$
$$u_3' = u_4$$
$$u_4' = -cu_4\sqrt{u_2^2 + u_4^2} - g.$$

For a given launch angle α, the initial conditions are known and so the system may be solved, for example using RK4.

However, we have two remaining difficulties; we do not know α and we do not know the "flight time" for which the solution must be computed.

Note that for a given α we could compute the solution until we have $x = u_1 > 1000$ using a small time step. Next, local interpolation (linear would suffice) could be used to estimate the height u_3 when $u_1 = 1000$. Let's denote this value of u_3 as $\phi(\alpha)$. So, ultimately, our objective is to then solve the nonlinear equation of one variable

$$\phi(\alpha) = 150,$$

which could be done for example, with the secant method.

Embedded within the proposed solution, this example illustrates the basic idea of *shooting methods* for the solution of a two-point boundary-value problem. *The idea is to embed the solution to a related initial value problem in an equation-solver.* The resulting equation is then solved so that the final solution to the initial-value problem also satisfies the boundary conditions. The specific details will of course vary according to the particular type of equation and the nature of the boundary conditions.

In what follows, the focus is on the solution of a second-order differential equation with boundary conditions which specify the values of the solution at two distinct points. That is, the problem is defined as

$$y'' = f(x, y, y') \tag{7.29}$$

subject to the *boundary conditions*

$$y(a) = y_a, \quad y(b) = y_b. \tag{7.30}$$

Under reasonable conditions, we expect the initial value problem

$$y'' = f(x, y, y'); \quad y(a) = y_a, y'(a) = z \tag{7.31}$$

to have a solution for each value of the *unknown* parameter z. Our objective is to find the appropriate value of z so that the solution of (7.31) hits the "target" value $y(b) = y_b$.

We need to express this as an equation for the unknown z. Let the solution of (7.31) for any value of z be denoted by $y(x; z)$ and define the function $F(z)$ by

$$F(z) = y(b; z) - y_b. \tag{7.32}$$

So our problem is to solve the equation

$$F(z) = 0 \tag{7.33}$$

for then

$$y(b; z) = y_b$$

and so the function $y(b; z)$ is a solution the differential equation (7.29) which satisfies the boundary conditions (7.30).

To this end, we have a nonlinear equation for z (7.33) that depends on output from differential equation solver. One approach is to use the secant method since it does not require any derivative information about F. The secant method requires two initial guesses (or estimates), let's call them z_0, z_1 for z. Then we would need to solve the initial value problem (7.31) for each of them. These yield $F(z_0)$ and $F(z_1)$ after which we can apply the secant method (Sect. 5.5) to generate our next estimate z_2. The secant iteration can then proceed as usual – the only difference being the need to solve a second-order initial-value problem on each iteration.

This approach exemplifies the need for efficient techniques for solving some of the fundamental problems arising in the real world. The need for an efficient equation solver is much easier to appreciate when evaluating the "function" entails the accurate solution of a system of differential equations (or in many real-world application, large systems of partial differential equations that comprise an off-the-shelf industrial simulation tool). Keep in mind, the underlying differential equation likely needs to be solved repeatedly for different initial conditions, meaning the need for efficient methods of solving differential equations becomes apparent as well.

The process is illustrated using examples below (the projectile motion application will be revisited in the exercises). The first iteration is described in considerable detail. Later we show how to set up the function F as a Python script so that the secant method can be used just as if it were a conventional function defined by some algebraic expression (but it is actually treated like a "black-box").

Example 11 Solve the Bessel equation $x^2 y'' + xy' + (x^2 - 1) y = 0$ with boundary conditions $y(1) = 1$, $y(15) = 0$ by the shooting method.

First note, this problem appears in Exercise 5 of Sect. 7.4. The solution of the initial value problem $x^2 y'' + xy' + (x^2 - 1) y = 0$ with $y(1) = 1$, $y'(1) = 0$ has a positive value for $y(15)$. This corresponds to $F(0)$ since the initial value $y'(1) = 0$ and, in this case, $F(z) = y(15, z) - 0$. Specifically with $N = 280$ steps, the following procedure would be used:

1. Rewrite the second order differential equation as a system of differential equations.
2. Use RK4 for systems with 280 steps between 0 and 15 with the initial conditions $[1, 0]$ (i.e $z_0 = 0$)
3. Set $F(z_0) = 0.2694$, the approximate solution to $y(15)$ (i.e. output of the previous step).
4. For a second value, take $z_1 = 1$ and use RK4 for systems with 280 steps between 0 and 15 with the initial conditions $[1, 1]$.
5. Set $F(z_1) = 0.5356$, the (new) approximate solution to $y(15)$ (which is actually worse than the first).
6. Calculate a secant iteration using

$$z_2 = z_1 - \frac{z_1 - z_0}{F(z_1) - F(z_0)} F(z_1)$$

$$= 1 - \left(\frac{1 - 0}{0.5356 - 0.2694} \right) 0.5356$$

$$= -1.012021$$

The next iteration begins with the solution of the differential equation with initial conditions $y(1) = 1$, $y'(1) = -1.012021$ which will give $f(z_2) = -9.5985 \times 10^{-5}$

Note that this is the solution to reasonable accuracy using just one iteration of the secant method. The two initial guesses and the final solution are plotted together in Fig. 7.7 along with the target point. The final curve agrees with the target and is marked with a '*'.

This example is misleadingly very successful, which is not always the case. For that example, Bessel's equation is a linear differential equation even though its *coefficients* are nonlinear functions of the independent variable. The general solution of such a differential equation is therefore a linear combination of two linearly independent solutions. (In this particular case, these are usually written as the Bessel functions of the first and second kinds J_1 and Y_1.) It follows that $y(15)$ is just a linear combination of $J_1(15)$ and $Y_1(15)$. Details of Bessel functions are not important to this argument, what is important is that the equation $F(z) = 0$ is then a *linear* equation. Since the secant method is based on linear interpolation, it will find the solution in only one iteration in this case.

Fig. 7.7 Illustration of the shooting method for Bessel's equation

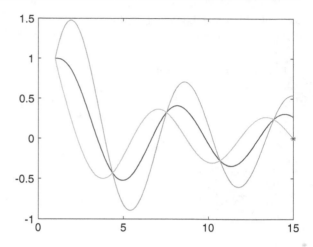

In general we can't expect to see such fast convergence. However, the procedure is basically the same and we demonstrate the algorithm further on the next example; a nonlinear differential equation.

Example 12 Solve the boundary value problem

$$y'' = 2yy'; \quad y(0) = 1, y(1) = -1$$

using the shooting method.

We take as initial guesses $y'(0) = 0$ which yields the solution $y(x) = 1$, and $y'(0) = -1$.

Note that for this problem,

$$F(z) = y(1; z) - (-1)$$

To get started, set the boundary condition to $[1, z_0]$ with $z_0 = 0$ and use RK4 for systems to solve the problem. This gives $F(Z_0) = 2$. Next, repeat the process with $z_1 = -1$ and get the approximated value for $y(200) \approx -0.6895$. This gives $F(z_1) = 0.3105$.

To complete the first secant iteration, we set

$$z_2 = z_1 - \frac{z_1 - z_0}{F(z_1) - F(z_0)} F(z_1) = -1.1838$$

The solution of the initial value problem corresponding to this initial slope is then computed.

This procedure can be automated by having a script that calculates $F(z)$ for any initial slope $y'(0) = z$. The steps used above show how to achieve that. Note that a function evaluation requires a call to the RK4sys solver. With that subroutine in hand, the regular secant method can be applied to the equation $F(z) = 0$, giving $z = -1.3821$. The actual error at the right-hand end of the interval is about 5×10^{-9}.

Note that the boundary conditions could be specified in a number of ways. For example if we had specified the *slope* of the solution at a different value, we would simply have compared the computed slope with that value.

Unfortunately, the method may be sensitive to the initial conditions. Recall the motivating example, finding the launch angle for a projectile to hit a specific target. The second-order air-resistance model the differential equations are a pair of nonlinear second-order equations which we can rewrite as a system of four equations. The major difference is that the boundary conditions specify a *position* whereas the independent variable is *time*. That is, the boundary conditions are given at an unspecified value of the independent variable. This problem can still be solved with the shooting method, but one needs to check the y-coordinate at the point where the x-coordinate is closest to the desired target and use that to measure how far we are off-target. See the exercises for more details.

Exercises

1. Solve the Bessel equation $x^2 y'' + x y' + (x^2 - 4) y = 0$ with the boundary conditions $y(1) = 1$, $y(10) = 1$.
2. Solve the Bessel equation $x^2 y'' + x y' + (x^2 - 4) y = 0$ with the boundary conditions $y(1) = 10$, $y'(10) = -1$. (This is similar to Exercise 1 except that the right-hand boundary condition is given in terms of the slope. Note that the other component of the solution to the system gives the values of the slope.)
3. Solve the equation $y'' = 2yy'$ with boundary conditions (a) $y(0) = 0$, $y(1) = 1$, (b) $y(0) = -1$, $y'(1) = 1$, (c) $y'(0) = 1$, $y(1) = 0$
4. A projectile is launched from the origin with initial speed 200 m/s. Find the appropriate launch angle so that the projectile hits its target which is at coordinates (100, 30). Assume the air resistance is proportional to the square of the speed with the constant 0.01 so that the differential equations are

$$x'' + 0.01 x' \sqrt{x'^2 + y'^2} = 0, \; y'' + 0.01 y' \sqrt{x'^2 + y'^2} = -g$$

For a given launch angle, compute the trajectory over a period of 10 s using a time-step of 0.1 s. Find the x-value in your solution that is closest to 100. For $h(\alpha)$ take the corresponding y-value and solve $h(\alpha) - 30 = 0$ by the secant method.

7.6 Conclusions and Connections: Differential Equations

What have you learned in Chap. 7? And where does it lead?

Many of the models that are derived from real-life physical situations result in the need to solve a differential equation. In most cases these will in fact be partial differential equations, or perhaps systems of ordinary differential equations. Often, they will have both spatial and temporal aspects which can result in computationally complex situations. To get an understanding of the basics of this enormous topic we begin with the simplest situation: a single first-order ordinary differential equation with a known initial condition.

Even for this situation, most practical examples do not admit paper and pencil solution by the methods of a typical first course in differential equations. The chapter began with Euler's method which provides a good introduction to most of the more advanced methods. Its essence can be summarized by "We are here, our current speed is 30 m/s, where will we be in 10 s?" The thing that separates this from a grade school word problem is the use of "current speed" implying that that speed (first derivative) is not constant, but instead depends on both independent and dependent variables.

Euler's method provides a basis for methods based on either a Taylor series or a numerical integration (Chap. 3) based theory which allows the development of higher order methods–meaning that the solution technique will be exact in the situation where the solution is a polynomial of higher degree.

The first higher order methods discussed here are the Runge–Kutta methods. Runge–Kutta methods are typically described as fixed step methods but often use intermediate points to gain higher-order behavior across that steplength. In numerical integration, we saw that accuracy can sometimes be enhanced by allowing different steplengths to be used in different regions. The same is true for Runge–Kutta methods and adaptive steplength methods such as Runge–Kutta-Fehlberg are widely used in practice–and are still being actively researched.

An alternative approach to higher order accuracy comes from multistep methods, which as the name suggests use data from multiple prior data points for explicit (Adams–Bashforth) methods in order to get a higher degree interpolant to the data. If the current target point is included in the calculation, the resulting Adams–Moulton method is called implicit. The two can be used to advantage as a predictor-corrector pair.

Systems of differential equations appear to pose a bigger threat but using the same approaches as already outlined but thinking of the independent and dependent variables as vector quantities allows the same methods to be extended fairly naturally. We also saw that higher order differential equations can be recast as systems of first-order equations and so the same techniques are applicable there, too.

There is however another big difference that can appear for higher order differential equations. Sometimes these are boundary value problems rather than the initial value problems we have discussed thus far. The simplest example is that of trying to hit a known target from a known initial point. That is we know the initial and final points but not full data to use the initial value methods. Shooting methods (so-called for obvious reasons) provide a good approach. We treat the second initial condition (typically the slope) as an unknown and then solve for that unknown such that the final point hits the target. The methods of Chap. 5 can be combined with our initial value problem methods to solve the resulting "equation"–one which we can certainly not expect to be able to write down explicitly.

In looking back at numerical integration and linear equations, you were actually already introduced to another approach to boundary value problems–explicitly the heat equation problems–using finite difference approaches to solve for all the intermediate points. This observation serves at least two purposes: one is that there are other approaches that we have not covered, and another is that the solution of differential equations is an appropriate final topic for this book because it relies on almost every topic we have studied.

Final topic, yes–but certainly not the end of the story! As well as more advanced study on the solution of ordinary differential equations in the manner we have discussed, the enormous issue of solving partial differential equations is the focus of whole journals on current research. Many of the approaches also utilize optimization techniques as well as all the topics we have discussed to address large complicated models. We do not even try to catalog major fields within that realm as they are simply too numerous! If you have been fascinated by some of what you've done in this course, you should certainly consider further study, possibly including graduate study as there is much important work still to be done.

7.7   Python Functions for Ordinary Differential Equations

Python has several functions available for solving ordinary differential equations through the scipy.integrate module.

scipy.integrate.RK23   is an explicit Runge–Kutta method using a second-order and third-order pair of Runge–Kutta formulas known as the Bogacki–Shamping formulas.

scipy.integrate.RK45   is an explicit Runge–Kutta method achieving higher accuracy than ode23 by using a pair of fourth-order and fifth-order Runge–Kutta formulas known as the Dormand–Prince formulas.

The two methods can be conveniently called through the common function scipy.integrate.solve_ivp. As an example, let us solve the equation from Example 2 equation $y' = (6x^2 - 1)y$ over the interval $[0, 1]$ for initial conditions $y(0) = 1$, using the RK23 method. We can do this using solve_ivp as follows

```
sol = solve_ivp(func, [0, 1], [1],
                      method = 'RK23')
```

where func is the function to be solved, defined as

```
def func(t, Y):
    return (6 * t**2 - 1) * Y[0]
```

Fig. 7.8 Plot of solution
obtained by the RK23 method

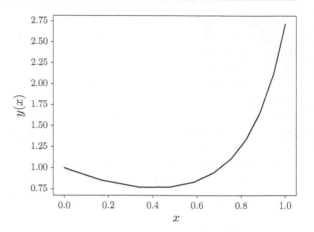

The output sol is an object containing various information about the solution. Most importantly, the fields t and y contain the time points used in the solution and the function values at these points, respectively.

There are optional arguments to this function to specify the time points at which to evaluate the solution and various other details.

We can plot the solution found in this example as follows

```
>>> plt.plot(sol.t, sol.y.reshape(-1), 'k-')
>>> plt.show()
```

generates the plot of y in Fig. 7.8. Note that the methods described here are from SciPy version 1.0.0. Previous versions of SciPy use an older interface through the function odeint.

Other functions   SciPy's integrate module has several other differential equation solvers. Particularly, the two above solution methods are recommended for "nonstiff" problems while two other methods Radau and BDF are intended for "stiff" problems. The method LSODA can handle automatic detection of stiffness and switching to an appropriate solver method. We have not defined what is meant by stiffness here and we do not cover the details. We encourage you to explore the SciPy documentation for further details.

Further Reading and Bibliography

1. Barlow JL, Bareiss EH (1985) On roundoff error distributions in floating-point and logarithmic arithmetic. Computing 34:325–347
2. Buchanan JL, Turner PR (1992) Numerical methods and analysis. McGraw-Hill, New York
3. Burden RL, Faires JD (1993) Numerical analysis, 5th edn. PWS-Kent, Boston
4. Cheney EW, Kincaid D (1994) Numerical mathematics and computing, 3rd edn. Brooks/Cole, Pacific Grove
5. Clenshaw CW, Curtis AR (1960) A method for numerical integration on an automatic computer. Numer Math 2:197–205
6. Clenshaw CW, Olver FWJ (1984) Beyond floating-point. J ACM 31:319–328
7. Davis PJ, Rabinowitz P (1984) Methods of numerical integration, 2nd edn. Academic Press, New York
8. Feldstein A, Goodman R (1982) Loss of significance in floating-point subtraction and addition. IEEE Trans Comput 31:328–335
9. Feldstein A, Turner PR (1986) Overflow, underflow and severe loss of precision in floating-point addition and subtraction. IMA J Num Anal 6:241–251
10. Hamming RW (1970) On the distribution of numbers. Bell Syst Tech J 49:1609–1625
11. Hecht E (2000) Physics: calculus, 2nd edn. Cengage, Boston
12. IEEE (2008) 754–2008 - IEEE standard for floating-point arithmetic. IEEE, New York
13. Kincaid D, Cheney EW (1991) Numerical analysis. Brooks/Cole, Pacific Grove
14. Knuth DE (1969) The art of computer programming, seminumerical algorithms, vol 2. Addison-Wesley, Reading
15. Langtangen HP (2016) A primer on scientific programming with python. Springer, Berlin
16. Munz P et al (2009) When zombies attack: mathematical modeling of an outbreak of zombie infection. In: J.M. infectious disease modelling research progress, Tchuenche and C. Chiyaka, pp 133–150
17. Olver FWJ (1978) A new approach to error arithmetic. SIAM J Num Anal 15:369–393
18. Schelin CW (1983) Calculator function approximation. Am Math Monthly 90:317–325
19. Skeel R (1992) Roundoff error and the patriot missile. SIAM News 25:11
20. Stewart J (2016) Calculus: early transdcendentals, 8th edn. Cengage
21. Turner PR (1982) The distribution of leading significant digits. IMA J. Num. Anal. 2:407–412

© Springer International Publishing AG, part of Springer Nature 2018
P. R. Turner et al., *Applied Scientific Computing*, Texts in Computer Science,
https://doi.org/10.1007/978-3-319-89575-8

22. Turner PR (1984) Further revelations on l.s.d. IMA J Num Anal 4:225–231
23. Volder J (1959) The CORDIC computing technique. IRE Trans Comput 8:330–334
24. Walther J (1971) A unified algorithm for elementary functions. AFIPS Conf Proc 38:379–385
25. Wilkinson JH (1963) Rounding errors in algebraic processes. Notes on Applied Science. HMSO, London

Index

© Springer International Publishing AG, part of Springer Nature 2018
P. R. Turner et al., *Applied Scientific Computing*, Texts in Computer Science,
https://doi.org/10.1007/978-3-319-89575-8